Planet on Fire

Planet on Fire

A Manifesto for the Age of Environmental Breakdown

MATHEW LAWRENCE AND LAURIE LAYBOURN-LANGTON

VERSO

London • New York

First published by Verso 2021
© Mathew Lawrence and Laurie Laybourn-Langton 2021

1 3 5 7 9 10 8 6 4 2

Verso
UK: 6 Meard Street, London W1F 0EG
US: 20 Jay Street, Suite 1010, Brooklyn, NY 11201
versobooks.com

Verso is the imprint of New Left Books

ISBN-13: 978-1-78873-877-4
SBN-13: 978-1-83976-491-2 (EXPORT)
ISBN-13: 978-1-78873-879-8 (US EBK)
ISBN-13: 978-1-78873-878-1 (UK EBK)

British Library Cataloguing in Publication Data
A catalogue record for this book is available from the British Library

Library of Congress Cataloging-in-Publication Data
A catalog record for this book is available from the Library of Congress

Typeset in Monotype Fournier by Hewer Text UK Ltd, Edinburgh
Printed in the UK by CPI Group (UK) Ltd, Croydon CR04YY

CONTENTS

INTRODUCTION

Not everything that is faced can be changed. But nothing can be changed until it is faced.

James Baldwin, 'As Much Truth
as One Can Bear', 1962

All states, markets, economies, welfare systems, militaries, major religions, scientific breakthroughs, cultures, medical advances, wars, and the people that fought them came about during a uniquely stable period of Earth's natural history. That era is now over. In its place comes a new era of accelerating environmental destabilisation. The accumulated burden of centuries of exploitation and destruction now threatens the safe functioning of the natural systems upon which all life ultimately depends. Ours is the age of environmental breakdown.

It is to a subset of this destruction that the world's attention has now turned: the accelerating crisis in the global climate system. The severity of the present situation was brought into sharp relief in October 2018 by a report from the UN's Intergovernmental Panel on Climate Change (IPCC), the global authority on the climate crisis. The report explored pathways to avoid 1.5°C of warming above the average temperature prevailing before the atmospheric onslaught unleashed by over two centuries of fossil fuel–powered industrialisation. It was the latest in a long line of increasingly stark, desperate warnings about the scale and severity of the climate crisis.

In the report is a graph – reproduced below – that illustrates how quickly greenhouse gas emissions must decline in order to stave off increasingly catastrophic climate breakdown.[1] Total global emissions must almost halve by 2030, and the global release of greenhouse gases must, in effect, end almost entirely by 2050. On the graph, these reductions are illustrated by a downward line, starting at the emissions rate around 2020 and falling to zero over the following thirty-odd years. The line – the pathway to a liveable and humane future – is so steep it is almost vertical. There are few, possibly no, historical examples of societies successfully undertaking such fundamental, transformative action in so little time.

This graph is one of the most important images in human history. It shows the existential challenge before us, the extraordinary changes required if life is to thrive.

The report received widespread coverage, its analysis summarised in a powerful, albeit misleading, message: we have twelve years to save the world. In truth, our planetary

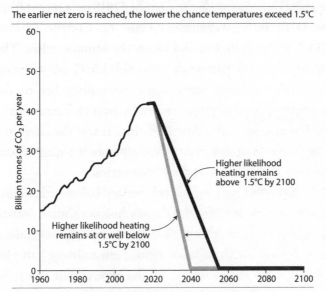

CO₂ emissions decline from 2020 to reach zero in 2055 or 2040

The earlier net zero is reached, the lower the chance temperatures exceed 1.5°C

Source: Valerie Manon-Delmotte et al., *An IPCC Special Report on the Impacts of Global Warming of 1.5°C* (2018)

emergency demands immediate and decisive action. Indeed, for hundreds of millions of people, the crisis is already here and has been for generations. And so it is not possible to draw a line dividing the binary between safety and catastrophe; but we can say that the faster emissions are reduced now, the less harm will be done in future.

However, even if half of all global emissions were eliminated by 2030, the prospect of climate breakdown would still not be averted. Net emissions must be reduced to and beyond zero, while dangerous tipping points in natural systems may have already been crossed. The violence of accelerating breakdown – falling disproportionately on those least responsible for the damage – is increasingly inscribed into our futures. The IPCC report was a clarion call in an age of profound emergency, a rallying cry for action soon to be amplified by campaigners around the world, from striking schoolchildren to Green New Dealers.

The Hour Is Late

The crisis is more severe than is generally realised. Warming is the result of the cumulative stock of emissions in the atmosphere. This has led the IPCC to set a 'budget' for the remaining emissions the world can afford to release if we are to have a fighting chance of avoiding temperature rises above 1.5°C. This budget, which underpins the 'twelve years left' story, is small. Smaller, for example, than the greenhouse gases set to be emitted by existing fossil fuel–dependent infrastructure – from cars to power plants – if they are used to the end of their productive life.[2]

Yet more emissions are on the way. When adding the contribution of fossil-fuelled power plants that are currently planned, licensed or under construction, let alone projected increases in car production, long-haul flights and carbon-intensive Western diets, emissions are set to far exceed that budget. It is estimated that Canada alone – where just 0.5 per cent of the world's population lives – could use nearly a third of the remaining global budget if it exploits all of Alberta's known tar sands reserves.[3] The result: the world is currently on track for increases in average global temperature of more than 3°C by 2100, risking globally catastrophic conditions in which, according to a research paper from JPMorgan Chase – one of the world's foremost funders of fossil fuels – 'human life as we know it is threatened'.[4]

Moreover, the proposed budget could itself be overgenerous. The IPCC's models assume the large-scale deployment, in coming decades, of 'negative emissions' technologies which would suck greenhouse gases out of the atmosphere, reducing their capacity to cause harm. But there is no silver bullet of this sort. Technologies are currently too expensive (potentially running into the hundreds of trillions of dollars), as yet untested or not even invented, or potentially devastating to ecosystems and societies.[5] Dangerously unregulated 'geoengineering' proposals, such as spraying reflective aerosols into the stratosphere, now litter the dreams of tech utopians.[6]

In other words, we are hurtling towards an unmanageable destabilisation of the climate system. Critically, however, the climate crisis is only one, if crucial, element of the wider systemic challenge. The total impact of human activity over many centuries has damaged the great natural cycles that underpin life on Earth to the point of breakdown. Soils are depleted, species are lost, water dries up. The cascading disintegration of the natural world will increasingly push societies to the brink of collapse. Environmental breakdown is not only about emaciated polar bears and plastic in the sea. It is about famine, drought and disease, and how their higher incidence could destabilise individual nations and entire regions, leading to social strife and war.

This Is about Power

Awareness of environmental breakdown is growing, even among mainstream voices and institutions. Environmental charities and corporate social responsibility 'first movers' are now joined by central bank governors, celebrities and media organisations all claiming to understand the crisis we face, and to be taking proportionate action. Technologists marvel at the falling prices of renewable energy. Governments commit to the successive phase-out of greenhouse gases, the banning of single-use plastics, and the investment of billions of dollars into the development of electric vehicles. Yet, overall, such efforts have thus far proved inadequate.

In the face of this failure comes all manner of handwringing, seeking to explain the absence of meaningful change. Often, the voices and institutions that make up the status quo fail to grasp, or wilfully ignore, the central insight needed to understand how environmental breakdown has come about, and what to do next: this is about power. It is about the power of extractive and exploitative economic structures – legal institutions, market arrangements, trade agreements and policy agendas – and the unequal dynamics they promote. It is about the central features of contemporary capitalism – the for-profit corporation, concentrated ownership, and over-mighty finance – that act as great engines of expansion and extraction. It is about the power of those states, companies and individuals that benefit from these arrangements, happy to see the world burn for the sake of maximising profit, maintaining the grotesque fiction that indefinite material growth is both possible and desirable.

This power is exercised with brutal impunity. Environmental organisations are banned, dismissed as terrorists and their protests suppressed. The plight of countries in the Global South torn apart by violent cyclones is roundly ignored by many in those Western nations whose emissions have supercharged these storms. Journalists are murdered while vast resources are lavished on misinformation, lobbying and nativist politicians. In turn, these politicians argue that environmental breakdown is a hoax or a minor issue, distracting us from policies that accelerate the destruction while demanding increasingly harsh responses to its impacts.

These varied evils are the common project of five centuries of global history. This is a story of extractive capitalism, of the intertwined histories of environmental degradation, colonial plunder, and power imbalances between Global North and Global South, the rich and the poor. It is precisely for this reason that environmental breakdown is an issue of justice and a reflection of the deep inequalities of power and wealth under capitalism. Those countries and communities who have contributed the least are already beginning to disproportionately experience the resulting suffering. Moreover, they often have a constrained capacity to respond, partly owing to the same destructive legacies of colonialism and unequal power dynamics that have generated the crisis.

Morbid Symptoms

In response to environmental breakdown and the deep inequalities within and between societies, morbid political symptoms are emerging. The exhausted political centre is floundering, lacking the ideas or political momentum to respond. The neoliberal status quo – that seeks to insulate capital from a democratic reordering – gestures towards sustainability, but without challenging the purpose and operation of the economy. Emerging from the contradictions and failures of the status quo comes a darker force: an explicit ethnonationalism that recognises the destabilisation to come and answers it with an appeal to protect privileged groups inside their societies and force greater pain on those outside. Nativist politicians in Europe falsely claim that 'borders are the environment's greatest ally' and 'it is through them that we will save the planet'.[7]

This is a politics that abandons a historical commitment to denialism and instead embraces growing destabilisation the better to realise a future of borders, stratification, and lethal inequality. Indeed, all manner of reactionary environmentalism grows from this destabilisation. Early in the Covid-19 outbreak, photos of deserted urban landscapes went viral with an accompanying slogan: 'Coronavirus is Earth's vaccine. We're the virus.'[8] Humanity was cast as the disease, a generalised plague on pristine nature to be controlled and managed; the sharp inequalities of power and unequal responsibility that have damaged natural systems were obscured and depoliticised. In a world of environmental breakdown, ethnonationalism sets us against one another, exploiting the economic and social pain of neoliberalism to redirect grievance at others. Theirs is a future of deepening eco-apartheid and social trauma.

Those seeking a better world will increasingly find themselves competing against this 'eco-ethnonationalism'. The outcome will decide how humanity fares under conditions of environmental breakdown. What comes next could be even more brutal and unstable. There is no guarantee of a better future: it is something that must be fought for.

The task is clear. We must rapidly and equitably transform the institutions, infrastructure, and ways of life that drive environmental breakdown, and make strides towards stability in little more than a decade. Doing so requires more than just a discrete series of policies to slow the destruction. We need a common project to transform our societies and support new ways of living and working. The purpose is not just to make today's economy environmentally sustainable but to build the democratic economy of tomorrow: dismantling the injustices of the present, replacing them with a reparative economy founded on the nurture of life, common care, and solidarity, enabled by institutions that share the wealth we create in common, and where meaningful freedom is a universal inheritance. A systems crisis must be met with a politics of systemic ambition. That is the central mantra of this book: against oligarchy and breakdown, we must build societies in which we can all flourish.

Beyond Barbarism: Towards Ecosocialism

In 1916, amid the wreckage of war and empire, Rosa Luxemburg saw that 'bourgeois society stands at the crossroads, either transition to Socialism or regression into Barbarism'.[9] Today, we once more stand at just such a crossroads: ecosocialism or barbarism. Yet this time really is different. Extractive capitalism tears apart the natural systems upon which all life depends and drives the disruption and violence of breakdown hurtling towards us. We are at a terminal juncture.

But we can still rescue our futures. Climate crisis and ecological breakdown are primarily a political crisis. We have the resources, technologies and ideas to decarbonise our economies and bring them within sustainable environmental limits; the challenge is mobilising the power and energy to match the scale of emergency, overcoming entrenched interests and inertia to drive transformation within a narrow time frame.

To do this, we need to replace the economics of extractivism with a twenty-first-century ecosocialism: the collective effort to democratise our economic and political institutions, repurposing them towards social wellbeing and individual flourishing, rooted in an abundant and thriving natural world. It is a goal that demands a different type of economy, one reoriented towards meeting social and environmental needs, overturning the injustices of contemporary societies and an extractive, neo-imperial global political economy; promoting communal luxury in societies of everyday beauty and comfort; expanding social ownership and control; and deepening and extending democracy and freedom. If you do not like the word 'ecosocialism', then use something else. But this is the systemic change we need to help us thrive as well as survive.

This is not a project of traditional statism, with success simply measured by the size of the state. It is about reimagining the foundational institutions of production, consumption and exchange, of work and leisure, based on principles of equality, deep freedom and collective empowerment, solidarity, sustainability and democracy. In place of the economics of enclosure and extraction, a twenty-first-century commons founded on collective stewardship; in place of concentrated economic power, a democratised marketplace where capital's monopoly over decision-making is replaced by social control and generative enterprise; against austerity, an ambitious mission-oriented state that lays the foundations for an expanded, decommodified public realm; and a reimagined household economy that dismantles unjust inequalities and increases leisure time.

Extending social control over finance to direct its power to serve real needs, valuing work that nurtures and sustains life, and democratising technological development and use to extend our capacity for creativity and mutual endeavour, ecosocialism builds the conditions for thriving. It works through a new internationalism, recognising that the failed age of the 'Washington Consensus' is over, and that, in its place, international cooperation must be built on a full understanding of how we got here and a shared view of where we go next. And it seeks to dismantle the structural racism and class inequalities that blight our societies.

This will require more than marginal adjustments. As signs at Black Lives Matter demonstrations across the world have spelled out, 'The system isn't broken, it was built this way.' To that end, we need one, two, many Green New Deals – adapted to local needs and cultures – confidently deploying the tools of public investment, democratic ownership, and green industrial strategy to rapidly remake our economies for sustainability and justice. The entwined environmental and inequality emergencies demand unparalleled ambition.[10] Corporate greenwashing, 'Green Deals' and limited 'green Keynesianism' won't suffice; the scale, complexity and compounding nature of the crisis demand a deeper, planned transition. An ambitious reordering of our economy is needed to secure more sustainable forms of abundance than the partial freedoms, inequalities and environmental crises of extractive capitalism.

What we propose is a politics for common care, mutual solidarity and joy, one that can renew hope against the deadening grip of the past: against the multiplying financial claims on our economic futures and the accumulated harms of environmental destruction that are now an existential threat to life. The environmental crisis changes everything: we can respond by delivering collective transformation.

Covid-19: A Transformative Crisis

If not now, when? The coronavirus pandemic, a profound public health emergency, has triggered the worst economic crisis in recent history. A viral cataclysm emerged from capitalism's entangled and devastating impact on the environment and spread through circuits of global production, falling upon healthcare systems weakened by a decade of austerity. In doing so, it was a warning from the future, a system-wide crisis whose exponential power overwhelmed the linear, outmoded politics of neoliberalism. It is the first crisis of the age of environmental breakdown.[11]

Covid-19 seized up the foundational mechanisms of capitalism; enforced isolation and social distancing wrenched relationships between capital and labour violently out of shape. The state re-emerged as an actor of unparalleled power. Value exchange was temporarily suspended and markets subordinated to the needs of public health.[12] An underlying crisis of care was sharpened. The uneven effects of Covid-19 once again exposed and deepened longstanding inequalities of gender, race and class. The virus may not discriminate, but our societies do, both structurally and systematically.

If the coronavirus crisis is a moment of deep trauma, it is also a potentially transformative juncture: old assumptions and settled conventions have been shaken, the limitations and inequalities of the status quo are exposed, and the work of producing and sustaining life, too often marginalised and undervalued, has been recognised and celebrated, rhetorically at least. As global financial capitalism stalled, and unprecedented public action to demobilise whole economies was undertaken, the boundaries of the possible were seemingly abolished. And into this historic moment erupted the global Black Lives Matter protests, a movement demanding justice long denied, seeking to dismantle interlocking forms of oppression, state violence, and the harms and inequalities of racialised capitalism.

All crises buckle and reshape the order of things; in what direction depends on politics and power. 'Crises are moments of potential change,' as Stuart Hall and Doreen Massey noted, 'but the nature of their resolution is not given.'[13] Even as the world fights the effects of Covid-19, a 'shock doctrine' politics of the Right is preparing to settle the crisis decisively in favour of the prevailing economic model: stripping back already limited environmental protections, further dismantling employment regulation, undertaking austerity 2.0. Against this, we must organise to win the crisis, building a politics of life from the worst public health emergency in a century.

If Covid-19 has buckled the pre-crisis status quo, then by definition a transformative response is needed, one that builds a new economy fit for human flourishing, rather than reinflating the inequalities and insecurities of the old. Emergency triage to stabilise the economy, while vital and necessary, is not enough. Creating a society of mutual care, solidarity and resilience will require the deep reimagining of the institutions and infrastructures of extractive capitalism. Severe as the crisis induced by Covid-19 may be, it is dwarfed by the twin crises of environmental breakdown and endemic inequality.

Winning Life

This book is an attempt to show what this transformation looks like and how it can be achieved, drawing on and complementing the wealth of publications, proposals and demands to have surfaced in recent months – and before. We have spent much of the last five years working in policy research and campaigning and found environmental breakdown was strikingly absent from mainstream policymaking. After the activism and political leadership of 2018 and 2019, this has begun to change. There are so many ideas and policy proposals, carried aloft by a diverse, powerful range of voices and movements. Yet even now, the scale and pace of environmental breakdown – and the actions needed to avoid the very worst – are strikingly absent from the policy agenda of governments around the world. This book is intended as a guide for understanding how we got here and exploring some of the ideas for where we go next.

While China, India, the EU and many others are critical actors in the struggle against environmental breakdown, our focus is primarily on the UK and US, the countries we know best, and which are central to the long story of environmental crisis. They both have a disproportionate responsibility for the problem and, to varying degrees, a capacity to take transformative action – as well as political movements eager to make it happen.

In the first three chapters, we set out how we got here: how an extractive model of capitalism has reshaped life on Earth, producing vast, unequal wealth, and unsustainable environmental and social harm. We set out the deep, systemic nature of the crisis; how it is broader, deeper and more dangerous than climate breakdown alone. And we make the case for a twenty-first-century ecosocialism, arguing it must first overcome neoliberalism to build an economy fit for life.

In the second half, we explore a programme of transformation. There is nothing inevitable about existing distributions of power and reward within the economy, nor necessarily a tension between environmental justice and sustainability and flourishing societies. We can organise the economy differently: through social control, not private dominion, through democracy, not oligarchy, and for environmental and social justice. But that will require a decisive break with the status quo. To that end, we explore how to democratise the state so it can steer the economy towards sustainability and justice; set out the contours of a new era of cooperation, from green community wealth-building to a new international order rooted in justice; a strategy for socialising finance so it serves people and planet; a radical expansion of the commons – not only nature, but also social and digital resources – to anchor communal luxury and a new era of public affluence; a new ecosystem of generative enterprise rooted in social ownership, purpose and control; and the reimagining of work around principles of solidarity and justice. This is a transformation to build economies that are democratic and sustainable by design. The concept of a Green New Deal sits at the heart of this transformation, and our primary focus is on its constituent components and its institutional enabling conditions.

However, a vision is not enough; we need a strategy to win. In recent years, the permafrost of 'no alternative' has thawed. Political parties and social movements at the heart of neoliberalism, the US and the UK especially, have belatedly put forward transformative programmes centred on the climate emergency; but nowhere have they won. That task increasingly falls to the millennial generation and those below it. Politics is a struggle between the competing claims of the past upon the present, and the ordering of the future by contemporary actions. Nowhere is this more true than in the case of environmental breakdown. If those leaders in older generations or in power refuse to help younger people realise a liveable future – as many have – they must step aside, before it is too late.

To strive for a liveable future for all, we must build on recent periods of advance, learn from missteps and defeats, and prepare the ground for a popular front capable of renewing economic and political hope.[14] As such, the final chapter charts a politics that can transform our societies. It speaks of the stories of radical hope we need to tell, of the coalitions we must build to drive deep change, and of the necessary antagonisms to overcome entrenched elites.

Against the deadening claims of an economics of no alternative, and the storms and environmental shocks, politics can reimagine many institutions falsely cast as 'natural' and permanent. The hopeful power of political life – of our collective capacity for renewal and world-making – can rescue our futures. This is a manifesto for a planet on fire – and a plan for life to flourish.

1

THIS IS ABOUT POWER

Whatever happens we have got
The Maxim gun, and they have not.
               Hilaire Belloc, 'The Modern Traveller'

Until the lions have their own historians, the history of the
hunt will always glorify the hunter.
               Chinua Achebe

In 1722, in a remote expanse of the Pacific Ocean, the Dutch explorer Jacob Roggeveen squinted out from the deck of his flagship at a triangle of land sitting on the horizon. Roggeveen had set sail nearly a year earlier to explore the uncharted seas west of South America. His aim: to find the mythic Terra Australis Incognita, the 'unknown land of the south', and map a western trade route to the lucrative spice markets of Southeast Asia. Rounding Cape Horn, his three ships spent nearly a month on the Juan

Fernández Islands, over 400 miles off the coast of Chile. It was here that Alexander Selkirk, the marooned Royal Navy officer who inspired the story of Robinson Crusoe, had been found a decade earlier. The isolation was profound. On 17 March they set off again, this time into the vast expanses of the Pacific. Nearly three weeks and over 1,500 miles later, they sighted an island on Easter Sunday and duly named it *Paasch-Eyland*, or Easter Island.

Roggeveen and his crew observed as few as 2,000 inhabitants occupying a tract of land notably empty of trees and studded with imposing *moai*, the monolithic head-and-torso statues carved from stone blocks erected to face inland, away from the sea. Roggeveen was astonished:

> These stone figures caused us to be filled with wonder, for we could not understand how it was possible that people who are destitute of heavy or thick timber, and also of stout cordage, out of which to construct gear, had been able to erect them; nevertheless some of these statues were a good 30 feet in height and broad in proportion.[1]

Subsequent visitors recorded as many as 900 *moai* across the island, over half of which remained mysteriously unfinished in the quarry. By the 1870s, as Western missionaries arrived, they found a population that had fallen to just over 100, barely surviving on the island's sixty-three square miles. How could this barren island have provided a home for such an extraordinary civilisation? What cataclysm had befallen the *moai*'s ingenious sculptors? These questions intrigued generations of Western explorers and scholars. John Linton Palmer – a Royal Navy surgeon on HMS *Topaze*, which, in 1868, stole the great Hoa Hakananai *moai* currently on display in the British Museum – thought the statues were 'made by a race passed away'.[2]

More recently, many scholars have posited that Polynesian explorers arrived around 800CE and began overexploiting its natural resources. Easter Island's extreme isolation meant its inhabitants were entirely dependent on the island's delicate ecosystem and natural resources, such as the thick forests of *Paschalococos disperta*, a native palm. Pollen analysis has shown that, by 1650, the Easter Island palm was extinct.[3] Evidence seemed to suggest that these trees were cleared to make way for agriculture and horticulture, to provide fuel for cooking, to build canoes, and to be used as rollers to move the *moai* from the quarry. Deforestation exacerbated soil erosion, impairing agricultural productivity, and overexploitation of land and sea birds drove a collapse in their populations. By around 1680, the inhabitants were reduced to burning grass and scraps of sugarcane for fuel, and had stopped building the *moai*. Environmental disaster soon graduated into social calamity. As the food ran out, the Rapa Nui's political order was overthrown by violent military leaders and the island fell into civil war. As society collapsed, desperate survivors turned to cannibalism.

By 1722, Roggeveen found a population that had dwindled from around 15,000 to as little as 2,000. Over the course of a few generations, a crescendo of self-inflicted environmental destruction drove the complex, thriving society of Easter Island to starvation, civil war and cannibalism. Jared Diamond featured this tragic tale in his bestselling book *Collapse*, presenting it as 'the clearest example of a society that destroyed itself by overexploiting its own resources.'[4] Writing elsewhere, he warned that

> Easter Island is Earth writ small. Today, again, a rising population confronts shrinking resources. We too have no emigration valve, because all human societies are linked by international transport, and we can no more escape into space than the Easter Islanders could flee into the ocean. If we continue to follow our present course, we shall have exhausted the world's major fisheries, tropical rain forests, fossil fuels, and much of our soil by the time my sons reach my current age.[5]

The complete eradication of the native palm was a crucial moment, depriving the islanders of fuel and, ultimately, their ability to grow food. This was, in effect, the islanders' choice, however subconscious. As one book on the Rapa Nui puts it:

> The person who felled the last tree could see that it was the last tree. But he (or she) still felled it. This is what is so worrying. Humankind's covetousness is boundless. Its selfishness appears to be genetically inborn. Selfishness leads to survival. Altruism leads to death. The selfish gene wins. But in a limited ecosystem,

selfishness leads to increasing population imbalance, population crash, and ultimately extinction.[6]

The Tragedy of Ecocide and the Mainstream Narrative

This telling of the history of Easter Island purports to show a simple truth. Societies that fail to respect environmental limits are doomed to commit 'ecocide', blindly exploiting nature to the point of collapse – and taking themselves with it. In many ways, this theory of environmentally induced suicide has come to dominate, explicitly or implicitly, the mainstream narrative of what has driven us to the point of disaster. In turn, it has a powerful effect on ideas for what we can and should do next.

A number of common themes are apparent. Human psychology sits at the forefront. In this interpretation, our evolutionary history predisposes us to selfishness and limits our ability to be mindful of the future. There was nothing particularly remarkable about the Rapa Nui; they were normal people looking out for their own interests and those of the people that mattered to them, much the same as us. And yet, even when it became apparent that the trees were at risk of disappearing altogether, the islanders still cut them down. This is the 'tragedy of the commons' at work: when resources are shared and those using them act according to their individual self-interest, the aggregate impact is to deplete the shared resource, damaging the common good.

Alongside innate psychology, this story asserts that three exacerbating factors drove the Rapa Nui to catastrophe: technology, population growth and the means whereby their society was organised. The island's first settlers brought the knowledge and tools with which they felled trees, hunted birds, caught fish, and erected the *moai*. The environment was impacted, but not critically, and prosperity came to the Rapa Nui. Families grew, society flourished and the population rapidly increased. More people meant more mouths to feed, more unthinkingly selfish humans stamping over a fragile environment, taking their impact to a critical scale. At this point, it could be that many of the islanders foresaw what was to come. But what chance did they have against the vested interests of the tribal chiefs, canoe makers and *moai* carvers? After all, those stripping back the Amazon today are 'only the latest in a long line . . . to cry, "Jobs over trees!"'[7]

In all, this version of the story of the Rapa Nui appears to teach that ordinary people and their societies are prone, by their very nature, to succumb to selfishness and short-term thinking. While the initial stages of environmental degradation may go unnoticed, a critical point is soon reached, at which the signs of disaster are unavoidable. At that moment, people can choose to indulge in selfishness and petty disagreement or to act, taking a more sustainable course. Ultimately, it comes down to choice. This is how the story of Easter Island is perceived to reflect our current moment. Scientists have told us we must rapidly reduce our environmental impact and yet people continue to fly, eat beef and vote for parties funded by fossil fuel companies. One day soon, someone will emit the last tonne of carbon permitted under the IPCC's carbon budget, or burn down the square mile of forest that finally triggers catastrophic dieback of the Amazon.

No wonder we find ourselves in a bind. While many governments promote clean technologies and pass regulations to slow environmental damage, they also subsidise fossil fuels and build roads, torn between environmental imperatives, the desires of voters, and the short-term needs of the economy. Campaigners do their best to shunt aside the vested interests of powerful companies and hold governments to account. But they struggle to engage the hearts and minds of most people, who would rather go on holiday and eat what they want. Every day, people the world over are doing 'their bit', recycling, voting, donating, campaigning, and making myriad consumer choices that add up to a prodigious whole. Yet what power do they enjoy compared to that of intransigent governments and wealthy companies? What difference do these individual choices make when coal power plants are still being built? Our mainstream eco-narrative finds itself mired in these tensions, stuck between the imperative to act and the inadequacy of its understanding of the causes of environmental breakdown. As disaster rages, the answer of the status quo is to accelerate or expand what it is already doing. More investment is needed, rapid deployment of increasingly brilliant technology, a greater purpose for businesses, persuasive campaigns reaching wider audiences, better consumer choices, changed voting patterns, new politicians, enforceable treaties, more facts, better arguments, less denial. For sure, more of all this is needed. But, as the clock ticks down, the storms grow, and birdsong fades, it is time to be honest about how it came to this. Let's start by revisiting the story of the Rapa Nui.

The Stolen Future of the Rapa Nui

There is a problem with the ecocide narrative of Easter Island: it is untrue.[8] Not only is this historically important, it also has profound implications for our contemporary understanding of environmental breakdown, and thus how we must act in the face of disaster. The real reason for the collapse of the Rapa Nui's society can be traced back to Jacob Roggeveen's voyage itself. One officer in his crew was struck by the island's lushness, describing it as 'a suitable and convenient place at which to obtain refreshment, as all the country is under cultivation and we saw in the distance whole tracts of woodland'.[9] The island's vibrant health at the time of European contact was further confirmed by a French visitor in 1786, who noticed a glaring contradiction with the stories of desolation disseminated by recent visitors:

> The aspect of the island is by no means so barren and disgusting as navigators have asserted . . . the accounts given of the inhabitants appear equally incorrect . . . Instead of meeting with men exhausted by famine . . . I found, on the contrary, a considerable population, with more beauty and grace than I afterwards met with in any other island; and a soil, which, with very little

labour, furnished excellent provisions, and in an abundance more than sufficient for the consumption of the inhabitants.[10]

While the island was almost completely deforested of native palm by the time of Europeans' arrival, this wasn't just the fault of the islanders. It was also a consequence of the voracious appetite of invasive rats, which had hitched a ride with the first colonisers.[11] Besides, the islanders used a variety of materials and techniques to cultivate gardens, fashion fish hooks and weave nets. Bountiful sea life was a primary source of nourishment and the islanders were observed as being 'exceedingly expert in the various methods of capturing them'.[12] Cultural practices placed limits on the exploitation of fisheries,[13] casting doubt on the accusation that the islanders suffered under the 'tragedy of the commons'. Thus the native palm was of little consequence to the Rapa Nui – including when it came to moving the *moai*, which, oral tradition suggests, were engineered to be 'walked' to their final destination, not pushed over wooden rollers.[14] In further contradiction to the mainstream narrative, evidence shows that the first people arrived in around 1200CE.[15] Over the following 500 years the population most likely increased to some 3,000 inhabitants, at which level, due to the inherent limits afforded by the environment, it remained. The islanders did fundamentally change the island's ecology, contributing to deforestation and other environmental degradation, erecting *moai*, and building homes and other structures. But many scholars of the island are today doubtful of the claim that the Rapa Nui's environmental heedlessness drove the collapse of the island's ecosystem.

A clue to what actually destroyed the island's habitat lies in Roggeveen's description of his landing:

> We marched forward a little . . . when, quite unexpectedly and to our great astonishment, four or five shots were heard in our rear, together with a vigorous shout of, 'it's time, it's time, fire!' On this, as in a moment, more than thirty shots were fired, and the Indians, being thereby amazed and scared, took to flight, leaving 10 or 12 dead, besides the wounded.[16]

While Roggeveen was able to establish peaceful contact with the Rapa Nui, this first, deadly contact with Europeans did not augur well for the future. Over the next 140 years, more than fifty European ships may have called at the island and there were many instances of European visitors seeing its people as 'sources of labour and, in the case of women, sexual satisfaction'.[17] Whalers would abduct islanders to supplement crews and practise their marksmanship by indiscriminately shooting indigenous people. By the 1830s, it was reported that sexually transmitted diseases had become a serious problem on the island.[18] Premeditated slave raids saw scores of men and women taken from the island. The Rapa Nui fought back, sometimes repelling invaders, but ultimately their efforts were futile. In October 1862 came the first in a series of devastating slave raids. By March 1863, over a thousand Rapa Nui had been captured by raiders and sold into slavery.[19] Soon after, a handful of surviving islanders were repatriated, bringing smallpox, which 'transformed the island into a vast charnel-house'. Following this, civil war and the collapse of the social order meant that 'when the first missionaries settled on the island, they found a culture in its death throes: the religious and social system had been destroyed and a leaden apathy weighed down the survivors from these disasters.'[20] In the 1870s, ownership of the island was seized by European traders, who set about forcibly removing the remaining population, burning huts and destroying crops to 'facilitate the persuasion of the starving natives who had thus little hope of surviving on their own island'.[21] In 1888, the island was annexed by Chile, of which it remains a part, and was almost entirely reserved for sheep farming by a Scottish company. By the end of the Second World War, the island was home to tens of thousands of sheep who irrevocably changed its ecosystem, transforming it into a vast, indistinct meadow; today, less than 8 per cent of Easter Island's wildlife consists of native species.[22] In 1966, the Rapa Nui were given Chilean citizenship. They are still, to this day, fighting for the right to their ancestral home.

Capitalism and Power, Coal and Slavery

The treatment of the Rapa Nui and the island at the hands of Western whalers, slave traders, missionaries and sheep farmers was described by Alfred Métraux, the eminent Swiss-Argentine ethnographer, as 'one of the most hideous atrocities committed by white men in the South Seas'.[23] A growing body of scientific evidence backs up this alternative narrative and a consensus is emerging that runs contrary to the ecocide hypothesis. In better understanding what drove the Rapa Nui to catastrophe, we find a story that is even more relevant to the current conjuncture.

History is littered with examples of misuse of local environments to the point of depletion or total destruction. Recent instances include overfishing, leading to the abrupt collapse of Atlantic cod populations in the early 1990s, and contemporary water crises afflicting cities across the world. In the extreme, this misuse could be a major factor contributing to the collapse of societies. Overexploitation of the 'carrying capacity' of local environments can exceed critical ecological limits. Fertile soils can be lost, rivers poisoned, fuel supplies exhausted. In turn, famine and strife can feed into existing political and economic instability, driving war and the breakdown of social institutions. This could be partly for the reasons featured in the orthodox story of the Rapa Nui: selfishness; short-termism; reckless use of damaging technologies; the pitfalls of how societies are organised and make decisions; the 'tragedy of the commons'. But it cannot be these alone. Other, more decisive factors are at play.

Fundamentally, environmental breakdown is about power. It is about who has it, how they got it, what they do with it – and what they do to hold on to it. It is about the power of economic systems and the dynamics they perpetuate. It is about the power these have over people, and the power of those actors – states, businesses, individuals – who profit from these arrangements. In the case of Easter Island, the desire for financial return drove raiders and traders deep into the empty stretches of the Pacific Ocean in the eighteenth and nineteenth centuries, at great risk to themselves and their crews. The bodies of the islanders were valuable, along with the island's fertile environment, because global systems of market exchange, of valuation and pricing, had deemed them so. In turn, the need to feed the insatiable process of compound capital accumulation compelled mariners to search every island, however remote, for cheap labour and natural resources. Legal systems, the cause and effect of cultural and religious convictions, dismissed the Rapa Nui as savages, legitimising their capture, sale, murder and rape.

If mainstream narratives accept that environmental breakdown is a problem, they often implicitly equate the world today to a planet-sized Easter Island, fearing that our inherent compulsion to commit ecocide is reaching a terminal phase, a 'tragedy of the commons' of global proportions. They rightly diagnose a number of factors that have brought us to this point but ignore others, brushing the inconvenient truths of imperial expansion, extractive capitalism and modern colonialism under the carpet. 'We're not wired to empathize with our descendants,' laments a psychology professor in the *Washington Post*.[24]

'We lack courage,' concludes President Macron to world leaders at the UN.[25] 'We need to do a much better job of informing people about the challenges,' warns Bill Gates.[26]

Thus the mainstream narrative is ultimately blind to the greater forces driving us towards global catastrophe: the compulsions of competition, compounding growth, and exploitation upon which extractive capitalism is founded. Without recognising these locomotive forces, the world is fated to bear witness to one failed summit after another, to watch greenhouse gas emissions rise and forests fall, and to absolve corporate leaders and racist politicians of responsibility, as societies slip into destabilisation and collapse. The stubbornness of the status quo in resisting the full connection of the past to the present is a disaster whose consequences we will all soon suffer.

The story of the two conflicting histories of Easter Island is itself an allegory for our age of environmental breakdown. On the one hand, the orthodox ecocide narrative is useful in reminding us that humans can and have always inflicted damage on the environment. But by limiting the scope of its analysis, it ignores the essential role of globalised profit-seeking and colonial expansion, erasing histories of environmental exploitation interwoven with human suffering and death that scar the identity of peoples to this day. In other words, far from being a neat fable of global ecocide, the orthodox story of Easter Island is a microcosm of the blind spots and shallow analyses of the popular understanding of the current crisis. Politicians and activist CEOs lament the global slide towards ecocide while failing to question the role of powerful financial firms and ignoring the displacement of peoples as ancestral lands are cleared to grow homogeneous biofuels, along with all the other symptoms of an exploitative economic system. Instead, we need a historical understanding of how profit, power, capital accumulation and colonialism binds the fate of the Rapa Nui to the desolation of South America in the seventeenth century, to the great furnaces of Victorian England, and to the global markets of the modern era.

A Short History of Environmental Breakdown

For the last 11,500 or so years, global environmental conditions have remained remarkably stable. This period has been dubbed the 'Holocene' by geologists, the keepers of the deep time of Earth's history. As the great glaciers of the Ice Age retreated, fertile agricultural land emerged and forests spread; then, as the climate calmed into balmy stability, the conditions for the flourishing of human society came into being. The Holocene epoch encompasses the entirety of written history, all technological revolutions and every major civilisation. Fast-forward to today and the growing crescendo of environmental breakdown heralds the end of the benign conditions of the Holocene, exposing the inadequacy of ideas and institutions forged in a more stable time.

Environmental destruction reached the global scale in the wake of decisive moments in the human story.[27] The first came as early humans left Africa, fanning out across continents. They arrived in the Middle East and Asia 100,000 years ago, Europe and Australasia some 50,000 years later, crossed into the Americas as little as 15,000 years ago, finally reaching New Zealand in around 1250CE. Wherever they went they encountered an abundance of species, large and small. These animals had had no previous contact with humans and therefore had little chance of survival as hunter-gatherers turned on them for food and clothing. In every part of the world, the arrival of humans led to a rapid reduction in fauna and a startling increase in species extinction. This process was accelerated by the emergence of agriculture, with farming gradually superseding hunting and gathering. Between 10,500 and 5,000 years ago, farming independently emerged in over ten locations across the world, from where it rapidly spread. As plants and animals were domesticated, two important things happened. Humans became increasingly locked into farming, which produced more food than hunting and gathering. More food led to more people, which in turn led to the need for more farming. Increasingly complex societies emerged, with priests, farmers and warriors all knitted together by the compulsion to cultivate and reap. Secondly, farming promoted certain plants and animals at the expense of others, forests were burned to clear land for crops, and more humans fanned out across the globe. So began an unbroken era in which humans increasingly pushed against the limits imposed by nature. Civilisations came and went but the global population inexorably rose, from less than 5 million at the advent of farming to over 400 million by 1500CE. Societies became more sophisticated and humans spread their environmental impact beyond borders. While these impacts were global in scope they had yet to become globalised, connected together by a web of economic systems amplifying their scale and pace, and had yet to cause damage of critical global significance, destabilising the Earth's life support systems.

By around 1300, the Medieval Warm Period – one of the Holocene's relatively minor temperature fluctuations – had created the conditions for an era of unprecedented prosperity. In Europe, the population had nearly tripled since the ninth century and farming increased sixfold, consuming the continent's ancient forests.[28] Feudalism reigned; societies were structured around a growing population of peasant cultivators who, in return for protection provided by a landholding elite, eked out a surplus from the land, which the elite frittered away on religious wars and shows of conspicuous consumption. When the Medieval Warm Period ended, a Great Famine began, as cold, wet weather decimated harvests; millions starved as landholders continued to demand a surplus. The weakened population was susceptible to the Black Death, which killed as many as half of Europe's inhabitants. Fewer peasants competing for the patronage of landlords eroded the power of the elite and efforts to re-establish the status quo triggered revolts that tore across the continent. Europe's crisis soon had implications for all the world.

From social collapse emerged economic dynamics of

global significance. As the old agrarian system melted away, custom and coercion ceased to become an effective means of enrichment for landowners. Instead, they had to turn to increasingly sophisticated markets and to employ a class of newly empowered tenant farmers. This arrangement incentivised both landowners and farmers to increase agricultural production and reduce the costs of doing so, in order to maximise profits. Investments were made in better agricultural tools and techniques, cheaper land and labour were sought, and the most lucrative crops, such as grain, were favoured. In all, a new economic system was forming around these dynamics, characterised by private ownership of and compounding investment in capital, with the prices, production and distribution of goods largely determined by marketplace competition instead of custom or birthright. In short, capitalism was arriving. With it came the compulsion to seek new frontiers, and new means of increasing production and reducing costs.[29] In England, this was partly manifested in landowners and farmers appropriating land previously held in common ownership, enclosing it for their private gain and using their influence over government to that end. More land was therefore given over to profit-making and private ownership, and farmers set about improving the means of getting more and more from the land.

These dynamics also compelled people to travel abroad in search of environments to exploit, better crops to grow, foreign markets through which to sell them, and, increasingly, the cheapest labour of all – slaves. All these factors combined to create the genocidal hell visited upon South America by Spanish colonisation during the 1500s. Western diseases and war ravaged the population within decades. Survivors were bound by systems of forced labour, which destroyed traditional ways of life, forcing indigenous communities into structures of work and exchange to earn the money needed to buy food and shelter. Plundered gold and silver, much of it mined by the local population, was used by the Spanish to fund wars of conquest and buy goods from the other side of the world. Environmental impacts had become truly globalised.

This happened through a three-step process of expansion, extraction and destruction. The compulsion to increase production and lower costs in order to maximise profit pushed the new economic dynamics across the world, connecting peoples and environments through globalised systems. These frontiers moved outwards, spreading capitalism over vast expanses of the world, and inwards, a fractal evolution of getting more for less from every field, forest and indigenous society these dynamics reached. In turn, wealth was extracted from places and peoples. Stories and legal structures grew up to justify the new state of affairs – demeaning the culture and lives of societies pressed into slavery, banishing their 'disposable people' outside of humanity, committing these 'simpler subjects' and 'lower races' to 'social death' – all while valorising the merchants and kings who benefited.[30] Then came the environmental destruction. Over the course of the sixteenth and seventeenth century, the scale and pace of human impacts on the environment exploded by an order of magnitude, fuelled by systematised human suffering. It took around 200 years to clear 12,000 hectares of forest in twelfth- and thirteenth-century France; by the mid-seventeenth century, colonisers in Brazil were using slave labour to clear the same area in a single year, energised by a booming market for sugar.[31] As the eighteenth century began, more European explorers, merchants and traders were getting in on the act. Increasingly complex financial structures sprang up to insure the ships, hedge against commodity price fluctuations, and provide loans to new companies. These included the British and Dutch East India companies, the first corporate giants, which turned their attention eastward. These connections of expansion, extraction and destruction soon metastasised into grotesque global webs, culminating in the transatlantic slave trade that intimately linked the fates of enslaved Africans and Easter Islanders, the ecological destruction of plantations in the West Indies, and the fancies of English consumer classes.

By the mid-eighteenth century, Britain had experienced generations of compounding investment in increasingly productive capital, while land enclosures had done away with subsistence farming, forcing the general population to rely on waged work and sprawling consumer markets for a living. These forces lit the first fires of what became the Industrial Revolution. Industrial technologies allowed European nations to harness the unprecedented stores of energy previously locked away in fossil fuels. Further technological development followed, enabling European ships, weapons and financial institutions to drive the next wave of expansion, extraction and destruction. Tentative trading operations became something more. In India, European hubs of exchange soon became the basis for the conquest of the entire subcontinent by private companies. Imperial rule spread across the globe and the map turned red, soaked in the blood of indigenous societies cut through by Maxim guns. Formal empires codified the process of extraction, with legal instruments condemning indigenous land as *terra nullius*. Great swathes of the world were opened up to mining and agriculture, a process dutifully enforced by rifles and gunboats. The environmental costs racked up and soon surged onto a momentous new frontier – the global atmosphere, into which industrial societies began to pour greenhouse gas emissions. The destruction had reached critical global significance.

On the eve of the First World War, a truly worldwide economic system had emerged. As John Maynard Keynes observed, this meant a wealthy Londoner in 1913 could

> order by telephone, sipping his morning tea in bed, the various products of the whole earth, in such quantity as he might see fit . . . The projects and politics of militarism and imperialism, of racial and cultural rivalries, of monopolies, restrictions, and exclusion, which were to play the serpent to this paradise, were little more than the amusements of his daily newspaper.[32]

The wealthy Londoner's enjoyment depended on a range of commodities, including rubber, which insulated the telephone cables and padded the wheels of delivery trucks. Much of this would have been extracted from the Congo, which suffered under murderous Belgian colonialism and where millions died under a regime that all too gladly fed the appetite of world markets for lucrative commodities. While Keynes's Londoner was lounging in bed, thousands of Congolese were, in the words of one local survivor,

> always in the forest to find the rubber vines, to go without food, and our women had to give up cultivating the fields and gardens. Then we starved ... When we failed and our rubber was short, the soldiers came to our towns and killed us. Many were shot, some had their ears cut off; others were tied up with ropes round their necks and taken away.[33]

The cataclysms of the First and Second World Wars saw a globalisation of coal and empire, underwritten by Western imperialism, evolve into a globalisation of oil and information technology entrenched through American hegemony. With it came the organisation of trading structures and governing institutions that bound the world ever more tightly into a global market system. Rapid technological innovation expanded the scope of extraction to every corner of the planet and drove a 'Great Acceleration' in population growth and consumption. In turn, environmental destruction reached exponential levels on a global scale, a daily process of desolation orders of magnitude above the gravest industrial disasters of the previous centuries. So was completed the human journey from resourceful ape to global force.

Past Is Prologue

The history of environmental breakdown is the story of the making of the modern world. It is a history that spans the emergence of capitalist economic structures, which drove dynamics of expansion, extraction and destruction as motors of human impacts on the environment. It follows their journey to globalising dominance, eventually wreaking destruction of critical global significance. In a matter of centuries, these structures and dynamics compelled more and more people, organised through increasingly complex states, companies and global markets, to expand the domain of their economic activity, to extract from peoples and nature, and to destroy them in the process. The true tragedy of the commons is that indigenous models of land stewardship that enabled generative coexistence with nature were brutally replaced by Western conceptions of ownership, their predecessors dismissed as savages and murdered. These stories are too often expunged from mainstream narratives of climate and other environmental breakdown, limiting our understanding of the deeper drivers of ecological destruction. Crucially, it is through remembering this history that we can recognise the path of dependencies determining the unequal impacts of environmental breakdown today. It is not the fault of the modern Rapa Nui that sea level rise is now eroding the island, shearing off cliffs, exposing the bones of their ancestors now lost to the ocean.

The inseparable histories of colonialism, economics and environmental breakdown have been told for centuries by those suffering on the front line. They have been largely ignored. More recently, geologists suggest that a new geological epoch is needed, to acknowledge that human-induced environmental change of planetary significance has ended the stable Holocene. Some argue for calling it the Anthropocene, an era defined by humans. But we are not all united as a singular humanity, or *anthropos*, in precipitating environmental breakdown and we cannot continue to adopt words and frames that occlude the story of how we got here. This is why some suggest that we now live in the Capitalocene[34] or the Plantationocene.[35] These alternatives remind us that without an understanding of economic history and its relevance to the current juncture, we cannot hope to grasp the inadequacy of ideas and institutions developed in a previous age, the age of the stable Holocene and that of colonial empire, a time of unsustainable resource use and global power imbalance.

But, before we choose an alternative path, we must first ask: how bad has it got?

2

FACING THE CRISIS

18:58 The people are hungry and thirsty, they had no water all day. The person on the phone is becoming difficult to understand as he loses strength. He asks for water, as a last resort. This situation is madness, we are being made to listen to them die of thirst!

Alarm Phone, an emergency hotline
for those crossing the Mediterranean

You have to understand,
that no one puts their children in a boat
unless the water is safer than the land.

Warsan Shire, 'Home'

In 1992, Henry W. Kendall, winner of the Nobel Prize in physics, wrote a letter. In it, he warned that 'human activities inflict harsh and often irreversible damage on the environment and on critical resources.'[1] If left unchecked,

Kendall wrote, 'many of our current practices put at serious risk the future'. He distributed the letter to fellow scientists around the world, over 1,700 of whom signed, including the majority of living recipients of all Nobel Prizes in the sciences. They were persuaded to do so by the overwhelming evidence that human destruction of the natural world had surpassed pressingly unsustainable levels.

They named the letter the 'World Scientists' Warning to Humanity'. It cautioned that the world would soon be 'trapped in spirals of environmental decline, poverty, and unrest, leading to social, economic, and environmental collapse'. In response, the scientists urged an end to environmentally destructive activities, the elimination of poverty, and the realisation of greater equality. It was decisive and accurate, the latest in a series of increasingly stark warnings from the scientific community, echoing centuries of concern from indigenous peoples and others on the front line of the war between economy and planet.

Twenty-five years later, in 2017, these warnings had largely gone unheeded. Global leaders – the CEOs, politicians, lawyers, financiers and others who control the great economic structures that govern our lives – had failed to address the spiral of environmental decline. Instead, environmental destruction had accelerated. This led 15,364 scientists from 184 countries to sign another letter, a 'second notice'.[2] In it, they highlighted the dizzying deterioration in many indicators of environmental health since the last letter, as shown in the figure below.

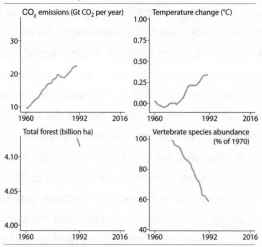

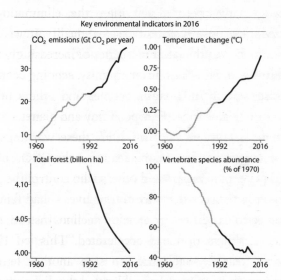

Source: Various , 'World Scientists' Warning to Humanity:
A Second Notice', *BioScience* 67, no. 12 (2017)

These letters teach us three important lessons. First, this isn't 'just' about climate breakdown. We face a far greater crisis. This is the age of environmental breakdown, of which the climate crisis is but a part, albeit an important one.[3] The natural world can no longer keep up with the speed at which resources are being consumed and ecosystems trashed: forests denuded, lakes dried up, soils stripped of fertility. The rate accelerates each year, far beyond the planet's capacity to regenerate, unleashing a destructive ferocity that now careers out of control.[4] Since the 1970s, nature has been damaged faster than at any point in human history, or, in some cases, millions or even billions of years. Extinction rates are many times higher than would normally be expected, rates unseen since the end of the dinosaurs.[5] The size of vertebrate populations – mammals, reptiles, amphibians, birds and fish – has dropped by an average of nearly two-thirds since 1970.[6] It is as if everyone in North America, South America, Africa, Europe, China and Oceania had died in the last fifty years.[7] Mass death exceeds 85 per cent in some countries and is worse for some species than others, with 83 per cent of freshwater vertebrates having died.

This is partly a consequence of how land is used. Animals are hunted to extinction, or their populations decimated by the loss of habitats. In all, three-quarters of the world's land, excluding areas covered by ice, is touched by humans, from sprawling cities, connected by oil pipelines, to vast industrial farms fed by brimming reservoirs.[8] As a result, great swathes of the Earth have been degraded, some critically.[9] Nearly 90 per cent of wetlands – marshes, swamps and mangroves that provide people and animals with homes and food – have been lost.[10] It is estimated that some 15 billion trees are felled each year; around half the world's forests have been cleared since the advent of human civilisation.[11] The soil problem is particularly worrying. In areas that are ploughed and tilled for farming, soil is disappearing over 100 times faster than it can be replenished by natural processes.[12] Nearly a third of the world's cropland may now have been lost, a global disaster.[13]

Soil is being degraded by a global food system hooked on fertiliser use and over-farming. This has affected the great global cycles of ecosystem fertility, such as the nitrogen cycle, which human activity may have changed more in the last 100 years than at any point in its 2.5 billion-year history.[14] In turn, pesticides are thought to be a major culprit behind the precipitous decline in insects observed in countries across the world. In 2017, a study in Germany recorded a 76 per cent loss since 1990.[15] Insects are a vital part of ecosystems, an essential link in all food chains, whose absence would see these chains unravel, threatening crops across the world.[16] Overall it is estimated that around a million species are threatened with extinction in the coming decades, an eighth of the total species on Earth.[17]

Crucially, too much damage and destabilisation can trigger 'tipping points'. These are rapid, violent and potentially irreversible changes in how natural systems work. Tipping points became relatively famous in 2018. As parts of the world burned and temperature records were shattered, the media picked up what might have been an obscure scientific paper warning of a 'Hothouse Earth' effect, in which the planet's overheating could trigger changes that further accelerate temperature rises.[18] Such processes include the thaw of polar permafrost, releasing the potent greenhouse gases trapped beneath, or the loss of Arctic summer sea ice that reflects warming sunlight back into space. The authors concluded that

> these tipping elements can potentially act like a row of dominoes. Once one is pushed over, it pushes Earth towards another. It may be very difficult or impossible to stop the whole row of dominoes from tumbling over. Places on Earth will become uninhabitable if 'Hothouse Earth' becomes the reality.[19]

A year later, fires tore through the Arctic Circle as the temperature in some areas topped 32°C, melting permafrost and destroying precious boreal forests at a rate unseen in ten millennia.[20] In turn, the fires released more greenhouse gas emissions than a large European country does in a year.[21] Similarly, in the Australian summer of 2019–20, bushfires released more CO_2 than the combined annual emissions of 116 countries.[22] They may have also killed or displaced some 3 billion animals. Ultimately, a tipping point in one natural system, such as the climate, can trigger rapid changes in another, leading to a cascade of environmental collapse as the delicate web of life comes apart.

Hence the second lesson provided by the scientists' letters: the destruction of the natural world has reached a terminal phase. Environmental degradation resulting from the combined impact of human activity has become a global force in its own right, threatening to destroy the natural foundations upon which we depend. Nature can no longer cope. Overall, the environment is being destabilised at such a scale and pace that the window of opportunity for avoiding some of the most catastrophic outcomes, including tipping points, is likely closing. In some areas it has already closed; great swathes of land have been exhausted and species are extinct. This is why the IPCC insists that global greenhouse gas emissions must be nearly halved by 2030, having peaked in 2020. It is why the UN secretary-general says that 'we face a direct existential threat.'[23]

The third lesson follows directly. Better evidence, clearer arguments, and more famous voices are not enough.

The Consequences of Environmental Breakdown

Climate breakdown is beginning to receive more media coverage in countries around the world. UN scientific reports grab headlines and campaigners flood the airwaves. This is a welcome development. But this coverage is not commensurate with what we face: a planetary catastrophe, beyond even the climate crisis. Popular comprehension of the consequences of environmental breakdown for societies is even less developed, on the whole, than our understanding of the problem itself. The focus is often on the fate of polar bears, elephants and other 'charismatic megafauna'; or on the beauty of nature and the aesthetic and psychological misery of its destruction; or on the localised impacts of environmental degradation, like harmful air pollution from cars, the perils of single-use plastics, or sea-level rise. These issues matter, and action is certainly needed to address them. But environmental breakdown transcends emaciated polar bears or plastics choking the sea. It is about famine, drought and disease, whose higher incidence will destabilise individual nations and entire regions, leading to social collapse and war. Incomplete or outright false political narratives belie the range and severity of negative impacts from environmental breakdown on societies around the world; they fail to tell us straight that, even with action, things will get far worse.

Nature isn't something that's just 'nice to have', treasured only for its aesthetic value; it provides the foundation upon which societies can exist, let alone flourish. The biological annihilation unleashed by economic systems is ultimately assaulting these foundations in a self-defeating spiral. This spiral has three stages. First, damaging environmental shocks occur in a particular place, affecting lives – human and non-human – and disrupting communities. The hit may be direct, occurring in the first instance as a result of environmental breakdown, such as when a storm destroys a city, maiming and killing its population and severely damaging its infrastructure. Direct impacts may also have indirect repercussions, such as economic and political chaos in the wake of a city-levelling storm. These direct and indirect impacts are already with us, and they are growing.

Second, the impacts are not confined to a single place but ripple out, transmitted and amplified through inter-connected social, economic and political systems. A good example is the food system. Food is often grown in one location, transported to another to be refined, processed and packaged, and then shipped to its final destination. Complex webs stretch across the world to serve a food system that ties countries to the fate of 'breadbasket' regions, in which the production of important global crops is concentrated. This system has been optimised to be effi-cient under conditions of environmental stability, which are now disappearing. A UK–US government food secu-rity taskforce has estimated that climate-induced disasters that formerly were expected around once a century could soon occur every thirty years.[24] The chance of a simultane-ous shock to breadbasket regions, such as the United States and China, which provide 60 per cent of the global maize supply, is growing.[25] The food system is not ready for this and so the impact on the global food supply and prices could be extreme.

This brings us to the third stage. As local impacts perco-late out through economic systems they interact with socie-ties of varying strengths and weaknesses, already experiencing a raft of problems. Countries and communities worldwide are dealing with the challenges of automation, rising geopo-litical tensions, population growth and the dynamic between older and younger cohorts, famine, obesity, social upheaval, economic downturns, and the enduring impact of the coronavirus pandemic. Environmental breakdown shakes this mix, supercharging and complicating societies' discontents. This spiral of destabilisation has often led environmental breakdown to be described as a 'threat multiplier'; it exacerbates existing instabilities across societies, connecting them together to create a system-wide storm of destabilisation. It comes as no surprise that military planners were one of the first groups to grasp the full implications. In 2014, for example, the US Department of Defense's Quadrennial Defense Review concluded that:

> The pressures caused by climate change will influence resource competition while placing additional burdens on economies, societies, and governance institutions around the world. These effects are threat multipliers that will aggravate stressors abroad such as poverty, environmental degradation, political instability, and social tensions – conditions that can enable terrorist activity and other forms of violence.[26]

In order to avoid overtly militarised language, let's use the term 'domain of risk' to describe how environmental breakdown is destabilising societies, from local to global levels.[27] This helps us set the challenge in historical context. There was a particular domain of risk during the Cold War, in which the simmering political and military stand-off between the United States and the Soviet Union underpinned a period of global geopolitical precariousness and localised hot wars, which included the possibility of global nuclear war – a risk that persists and is evolving. Another domain of risk was marked by the collapse of the post-war global financial order – the so-called Bretton Woods system – after which financial crises became more frequent and severe.

The New Domain of Risk

The domain of risk imposed by environmental breakdown is more serious than any previous example. It is largely driven by natural systems, which are global, highly complex, and, to a large extent, beyond our control. No area of society is left untouched, from local communities to international institutions. Furthermore, the risks are varied. They include catastrophic shocks that are hard to anticipate precisely, such as super-storms, pandemics or the collapse of a food chain. These shocks can strike in the midst of grinding, 'slow burn' crises; for example, the ongoing loss of food production in a region already experiencing economic hardship and political instability, which drives forced migration and political and economic problems, further destabilising the region. In this way, the risks imposed by environmental breakdown touch on everything, in an unprecedented, system-wide state of saturating destabilisation.

At this point, it's crucial to understand that environmental breakdown will inevitably get worse, as a result of the previous degradation and that which is still to come. In the case of climate breakdown, current temperature rises are the result of historical emissions and spark processes, like the melting of ice caps, that are slow to reverse. Even if Donald Trump had become a climate angel and Jair Bolsonaro dedicated his life to preserving biodiversity, we would still be set for worsening conditions, compounding the damage already underway.

Ultimately, the destabilisation unleashed by environmental breakdown could even trigger the collapse of social and economic systems. This can happen at the local level, such as the temporary collapse of economies in cities destroyed by supercharged cyclones. It's in these situations that 'disaster capitalists' pop up, finding new sources of accumulation in chaos and dislocation. We saw this when they came for Barbuda in 2017, after Hurricane Irma crashed into the island, using the evacuation of its population as a cover for bulldozing broken homes, and felling an ancient forest in order to build an international airport.[28] This is not only a new extreme normal, it is no normal: a constantly evolving state of complexity and uncertainty unprecedented in human history.

This Is about Injustice

In the final analysis, environmental breakdown is an issue of justice. The great systems of profit and power that have driven environmental breakdown also created a world of unjust structures that perpetuate vast inequalities. Those who contributed least to the problem suffer the brunt of its impacts. In the case of climate breakdown, low-income nations have made a negligible contribution to cumulative greenhouse gas emissions. From the mid-1700s, as proto-industrial fires raged and the Royal Navy protected the slave trade, Britain was a lone first in the international league table of annual greenhouse gas emissions. By 1850 it was the third-largest emitter, as the US and the rest of Europe caught up, and it now sits at around number fifteen. Meanwhile, the US has alternated between first and second place, having been recently displaced by China. In contrast, Ghana dropped from 103rd in 1850 to 111th in 2014, while Haiti climbed from 147th to 142nd over the same period.[29] In all, it is estimated that the poorest half of the global population contribute 10 per cent of yearly emissions, while the richest 10 per cent contribute half.[30] As such, most cumulative greenhouse gas emissions since the beginning of the Industrial Age have come from a few wealthy Western nations. This is also the case for other areas of environmental impact, with many European nations estimated to have an ecological footprint exceeding sustainable limits by over 150 per cent.[31] This inequality in contribution exists within countries. In the US, it is estimated that the top 10 per cent of earners contribute more than six times the amount of emissions than the bottom 50 per cent, with similar ratios seen in countries around the world.[32]

Those communities contributing least to environmental breakdown are often the most exposed and most vulnerable to its effects. Many of the world's poorest countries are located in regions suffering high temperature extremes and other environmental problems, and are already burdened by acute social and economic problems, something that equally holds for communities within countries. In many cases, those countries impacted the most are also scarred by the extractive legacy of colonialism. As storms tear through cities in the Global South, problems of governance and poverty can easily be traced back to the exploitative regimes of the Western nations whose emissions now alter global storm patterns. The extraction continues, as many nations are burdened with predatory loans and a multilateral system skewed towards corporate interests, factors which hamper the ability of these nations to prepare for and respond to environmental shocks. Meanwhile, the richest are better able to mitigate the impacts of environmental breakdown, partly because they have reaped huge economic benefits from the release of greenhouse gases, wholesale environmental destruction, and a leading, often extractive role in the global economy.

Dimensions of injustice intersect across ethnicity, gender, class and age, exacerbating deep structural inequalities. Younger generations and those not yet born have contributed little to a problem they will experience more than older generations.[33] Some of the world's youngest populations live in countries, particularly across the African continent, that are most severely impacted and are least responsible and less prepared for environmental breakdown. Women, already disproportionately suffering violence and oppression, are more likely to be affected, since the burden of natural disasters, food insecurity and pollution often fall harder on women than men.[34] The same can be said for ethnicity, with people of colour disproportionately bearing the brunt of environmental breakdown within and across countries.[35] These inequalities are inseparable from the structural injustices hardwired into social and economic systems. It is not mere statistical chance that sees certain ethnic, gender or income groups hardest hit by environmental breakdown. In some parts of the world, this is a result of explicit discrimination, as in the case of the systematic oppression of indigenous communities in Brazil and the destruction of ancestral lands and the precious ecosystems held within. It is also the result of the enduring legacy of colonialism and ongoing state-sponsored oppression, which are imprinted on cultural, educational, economic and political structures around the world, narrowing the prospects of people to this day. These structures entrench power imbalances that allow, for example, large corporations to exploit communities who struggle to make their voice heard, dumping waste and ignoring the mounting health costs of environmental degradation.

It is precisely because environmental breakdown is a problem that exists between and within generations that nations and peoples must take responsibility for their historical actions, particularly those that have disproportionately contributed to both cumulative degradation and the creation of structures of inequality. Failing such an approach, we speed towards a world of 'climate apartheid' in which the wealthy can pay to escape the ravages of environmental breakdown, leaving the rest of the world to suffer.[36] Recognising and acting on injustice is a political and moral imperative, as environmental breakdown can pit us against each other, gnawing at the bonds of our mutual humanity.

This is already happening. Western nations urge environmental restraint from former colonies while ramping up consumption at home and outsourcing pollution abroad. Politicians downplay or ignore the historical path dependencies that expose certain ethnic groups to higher levels of pollution. Billionaires build 'bolt holes' into which they can retreat when global catastrophe arrives. Younger generations, already struggling with high unemployment, expensive rents and excessive debt, look on, aghast at the grotesque spectacle of baby-boomer politicians investing in fossil fuel power plants. It is to the siren calls of these divisions that far-right movements, on the rise the world over, have begun to turn.

The Regressive Politics of Environmental Breakdown

On 3 August 2019, Patrick Crusius, a twenty-one-year-old Texan resident, uploaded a PDF to an online forum. In it, he had written:

> Our lifestyle is destroying the environment of our country. The decimation of the environment is creating a massive burden for future generations . . . This has been a problem for decades . . . Everything I have seen and heard in my short life has led me to believe

that the average American isn't willing to change their lifestyle, even if the changes only cause a slight inconvenience . . . So the next logical step is to decrease the number of people in America using resources. If we can get rid of enough people, then our way of life can become more sustainable.

Twenty-seven minutes later, Crusius started shooting people in a Walmart carpark in El Paso. Within moments, twenty people were dead. Crusius sought to target people of colour, writing that the attack was in 'response to the Hispanic invasion of Texas'.[37] The shooting was the latest in a series of mass murders perpetrated by men who claimed to be reacting to the 'great replacement' of white Western populations: the conspiracy theory holding that mass migration from abroad and a lower birth rate at home is leading to the 'genocide' of white populations. Their actions have been applauded by far-right users of online chat forums, 'ethnonationalists' emboldened by the success of like-minded politicians who brandish concepts of racial and national identity. These mass murderers and their online cheerleaders are increasingly linking racist, anti-immigrant dogma to environmental concerns. Their narrative is simple: resources are scarce and environmental breakdown is increasingly robbing us of the beauty and benefits of nature – something impossible to reverse if countries welcome growing numbers of climate refugees. Of course, the immigrants are rarely white, and the issue is never the consumption habits of wealthy nations and individuals; everything often comes down to population growth in the Global South.

These developments expose how environmental campaigning tends to rest on two complacent assumptions. First, regressive forces, including intransigent right-wing politicians and their corporate sponsors, are only capable of denying the problem outright or delaying measures to realise a more sustainable world. Second, progressive movements seeking the opposite ends are on the side of the angels, equipped with the science, ideas and will to prevail. As school strikers, activist CEOs and enlightened media organisations join forces with veteran environmentalists, social justice campaigners and indigenous communities, these two assumptions lay the foundation of a seductive narrative. Let's call it the 'good always wins' story. It goes something like this. Progressive forces are locked in a battle against regressive deniers and delayers. Progressives will ultimately triumph as more and more people see the light, convinced by the visible horrors of environmental breakdown or won over by the promise of a better world. As new governments are elected and outdated social, economic and political systems are reformed, the will of the people will finally be realised, vanquishing greedy corporations and their science-denying political lackeys, averting catastrophe in the nick of time.

There are three major problems with the 'good always wins' story.

First, it is not only those identifying with progressive causes that see both threat and opportunity in environmental breakdown. This is a problem caused by economic systems and the legacy of colonialism, by short-term profit-making for the few over long-term flourishing for the many. For their chief beneficiaries, any threat to these systems must be rebuffed, and so vast resources are lavished on misinformation campaigns, think tanks and politicians, with considerable success. Meanwhile the chaos unleashed by storms and desertification presents new frontiers for extraction and exploitation, as disaster capitalists race to build casinos over cyclone-wrecked homes and buy financial instruments that bet against bountiful harvests. As breakdown and disruption spread more people will join the ranks of the displaced, dispossessed and downright scared.

Second, while progressives have traditionally appealed to science and rationalism, making the case for action through clear, evidence-based argumentation, resurgent regressive forces have proved adept at appealing to emotion. In a world in which fears over immigration are already stoked and exploited for nativist ends, it may be that the invocation of 'mass migration' coming from countries crushed by climate breakdown could be a more politically successful narrative than one based on science and rationality.

The third point follows directly. There is a regressive politics of environmental breakdown, and it could win big as things continue to get worse. If this politics denies or delays, it does so to allow corporate giants to continue to plunder the Earth with impunity, or out of fear of a popular backlash against green measures. If it acts, it does so too slowly, without threatening the fundamentals of the great dysfunctional system over which it presides, placing the burden on those who can bear it the least. It pits people's present dependence on this system against their mounting concern for the future. As the seas rise and land depletes, it uses anxiety and fear to mask profitable exploitation of the chaos, recasting the suffering of others as a threat to 'the people'. In turn, it advocates the harshest possible response to the destabilisation brought by environmental break-down, intensifying a violent regime of walls, naval block-ades, and detention centres, eroding basic liberties and normalising the creep of authoritarianism.

The Threat of Eco-Ethnonationalism

In many respects, we already live in a world in which a regressive politics of environmental breakdown reigns. The elements of this political project fall into three distinct strands: Status Quo Neoliberals, Denialist Conservatives and outright Eco-ethnonationalists.

Status Quo Neoliberals enjoy power across the world as the mainstream defenders of the financialised, privatising model of capitalism that emerged in the late 1970s, despite struggling under the weight of its own contradictions since the financial crisis of 2007–08. While maintaining other-wise, they are selective in their espousal of the science, focusing almost exclusively on the climate emergency. They have been told by scientists and activists that we can only avoid catastrophic environmental breakdown by real-ising a 'fundamental, system-wide reorganization across technological, economic and social factors, including paradigms, goals and values', as a 2019 UN report concluded.[38] Yet they reject systemic change, advocating limited, market-based solutions. Where progress is made, it is slow and largely focused on greenhouse gas emissions. These neoliberals do not seek to meaningfully address injustice, placing the burden of policies on poorer groups and imposing austerity policies while giving tax cuts to the wealthy. This was the case in France, when Emmanuel Macron increased fuel duty in 2018 as part of a range of fiscal reforms that disproportionately harmed low-income groups, while benefiting the top 1 per cent of earners.[39] Status Quo Neoliberals act to maintain the structures of a global economic system that actually relies on environmental breakdown and its inequitable impacts. They fail to meet their commitments under the international environmental treaties which they claim to valorise, while pumping public money into fossil fuel investments in the Global South. As environmental breakdown worsens, they continue to advocate for the militarisation of borders amid a sweeping anti-immigration rhetoric.

Denialist Conservatives, for their part, either cast false doubt on the science or explicitly deny it. They differ from many Status Quo Neoliberals by eschewing liberal internationalism, protesting or threatening to withdraw from environmental treaties and seeing the world as a zero-sum battle of all against all for finite resources. Albeit proclaiming themselves the enemy of corrupt elites, they actively promote the project of Status Quo Neoliberals, investing in fossil fuels, accelerating deregulation, cutting taxes for the rich, deepening financialisation and eroding state capacity. When such processes intensify popular resentment at the broken promises of the prevailing economic system, they shift the blame onto immigrants, ethnic minorities, and the impacts of trade and foreign competition. Donald Trump is the archetypal Denialist Conservative. But he also epitomises the way denial has always been complemented by an implicit acceptance of environmental breakdown. For example, even as the Trump administration was withdrawing from the Paris Agreement, it ramped up American strategic interests in the polar regions with a view to exploiting the opportunities opened up by melting ice.

In many ways, Denialist Conservatism has begun to morph into explicit Eco-ethnonationalism. Until the late 2010s, many far-right nativist parties were united in denying environmental science. Now, many of these parties are tentatively embracing the reality of breakdown, using it to argue that environmental crisis follows directly from (non-Western) overpopulation and multicultural globalisation, and so immigration must be stopped to preserve the environment. In 2019 the French politician Marine Le Pen, head of the Rassemblement National (previously the Front National), said that one 'who [is] rooted in their home is an ecologist', in contrast to those who are 'nomadic . . . [who] do not care about the environment; they have no homeland'.[40] In the European Parliament elections of that year, a leading party figure affirmed: 'Borders are the environment's greatest ally; it is through them that we will save the planet.'[41] In the face of environmental destabilisation, Eco-ethnonationalists employ 'lifeboat ethics', arguing, as one influential far-right thinker has written, that 'when the lifeboat is full, those who hate life will try to load it with more people and sink the lot. Those who love and respect life will take the ship's axe and sever the extra hands that cling to the sides.'[42]

These ideas are reaching a younger audience through slick content promoted by 'identitarian' activists who seek white ethno-states, urging them to 'plant more trees, save the seas, deport refugees', in the words of one viral sticker campaign.[43] It is out of this melting pot of fear and loathing that Crucius and other racist, misogynistic mass killers have emerged.

These three political currents are not interchangeable, and enjoy varying degrees of power. But all three share common features: they favour powerful vested interests that profit from breakdown, they seek to preserve key economic structures antithetical to environmental stability, and their actions amplify social and economic injustice. In this way, these three currents could be mutually reinforcing as environmental breakdown accelerates and societies come under mounting stress. The UN and World Bank have, for example, warned that environmental breakdown drives forced migration and could increase refugee numbers by over 100 million by 2050.[44] This is on top of the tens of millions already displaced by conflict and persecution as well as by environmental disaster. While the vast majority of forced migration occurs locally, trapping people within the borders of a country or region, politicians from the three currents have, to varying degrees, proven adept at whipping up fears in response to the 'migrant crises' of the 2010s. These narratives are already evolving to link forced migration to environmental breakdown. In 2018, nativist European parties protested a non-binding UN migration compact that recognised climate breakdown as a driver of forced migration, because it risked opening a 'Pandora's box': climate breakdown would become a 'recognised justification for asylum', leading Europe to be 'overrun by millions of climate refugees'.[45]

As the world suffers under increasingly severe environmental breakdown, these reactionary political currents will doubtless wheel out the supposed threat of mass migration. It will rest on racist stereotypes and ignore the responsibilities of wealthy nations and the pain of those who have been displaced. It will be used to legitimise militarisation and growing violence, under cover of the Trumpian promise that it's not so much about keeping them out as about keeping us safe. This is already happening. Although Macron urged action to 'make the planet great again', he also supported the 'externalisation' doctrine of Frontex, the European border force, which de-prioritises the protection of life. While the British Conservative government brags about its climate credentials, it also dispatched the Royal Navy to intercept a few small boats of people adrift in the English Channel over Christmas 2018, declaring it a 'major incident'.

Under conditions of environmental breakdown, the shared traits of Status Quo Neoliberals and Denialist

Conservatives are uniquely suited to push both tribes towards elements of an explicitly Eco-ethnonationalist agenda. Destabilisation and collapse play on their worst traits, in a mutually reinforcing process of convergence towards a violent zero-sum nativism as fires rage and crops fail. It could come in an unbroken line, traced from the dehumanisation of modern capitalism, through Trump and Bolsonaro, onto their heirs. Or it could be born from the failings of the 'good always wins' story, with progressive movements caught off guard by a rapid succession of environmental shocks that overwhelm the administrations of President Ocasio-Cortez and like-minded allies. In every country around the world, Millennial Trumps and Bolsonaros can confidently look to the future and be assured that the conditions for their success are growing. Pain leads to grievance, which is redirected at others. Promises of material consumption and status draw a critical mass of continued support. Soon, great reservoirs of hate are allowed to breach and any action to stem environmental breakdown is abandoned, dismissed as a luxurious distraction from the necessity of national protection. Eventually, the global good of cooperation is crowded out by domestic stresses, leading to a breakdown of coordination as polities turn inward, back to the protection of the nation state, or against each other. This is what global collapse looks like.

As things stand, there is too great a risk that a self-defeating politics marrying inadequate action and violent reaction is set to decide the course of history in the era of environmental breakdown. We need an alternative.

3
BEYOND THE RUINS

People are suffering. People are dying. Entire ecosystems are collapsing. We are in the beginning of a mass extinction, and all you can talk about is money and fairy tales of eternal economic growth? How dare you?

Greta Thunberg, speech at the UN
(September 2019)

Reclaiming our future from conditions of accelerating environmental breakdown requires us to act on the central lesson of the history that has brought us to this point. This is about power: who has it, how they got it, how they exercise it. It is about who owns what, who makes the rules and for whose benefit, how cooperation is organised, and how costs and rewards are distributed.[1] The way that this power is constituted and used is driving breakdown. This state of affairs is determined by the prevailing political-economic paradigm: the dominant set of theories and assumptions, goals, institutions, policies and narratives that create, justify and perpetuate structures of power across and between societies.[2]

Human societies have undergone extraordinary change in the past 500 years, and with them, their governing ideologies and paradigms. Global market systems grew out of pre-colonial privateering and plunder. Formal empires fell and welfare states emerged. Total war created vast state apparatuses that were repurposed to encourage and protect the power of a global finance system, which has metastasised to become society's master. Yet throughout these shifts, and even as the dominant ideas, goals, policies and stories determining the structure of our societies have transformed, one constant has remained: the externalisation of nature from most human societies, its brutal transformation into unequally shared wealth, and the organising of the environment to benefit human hierarchies of wealth and power, regardless of the consequences for the stability of natural systems.

The result: countries across the world have enjoyed profound material progress, but at the cost of environmental stability. This has resulted in a positive correlation between progress towards social goals and environmental breakdown: the better social needs are met, the more the environment is destroyed. In this way, no country can be rightly seen as a 'developed nation', because none has yet managed to provide a decent life to its inhabitants without destroying the natural preconditions upon which society depends. We all live in developing nations.[3]

The Tyranny of No Alternative

The dominant political-economic paradigm is neoliberalism, the latest evolution of the economic structures and dynamics that have driven environmental destruction over the last many centuries. Neoliberalism is many things: an often-contradictory strategy for regulating capitalism; a strong rather than small state, used to enforce market-based forms of measurement and evaluation into ever more domains of life; a mode of governance and rationality; and an ideological and class project that extracts wealth and power upwards. At its core, it is a political project to insulate capitalism from democracy, to transform the economy into an object beyond the realm of politics, rendering the 'market' and unequal forms of economic power safe from democratic intervention.[4] This has resulted in a number of consistencies in neoliberal statecraft, even as its specific institutional arrangements have varied over time and place: the dismantling of the power of organised labour; GDP growth as the lodestar of economic policy, regardless of how it is distributed or the externalities it generates; active support for a deregulated financial system, which is nonetheless propped up in times of crisis; the extension of market relations, measurements and subjects in place of a universal welfare state and the political allocation of goods and services; the facilitation of monopoly and rentier power; and the privileging of private wealth over social affluence. This assemblage of policies has vital consequences for how our relationship with the environment is organised, by whom and in whose interest.

The logic and structures of neoliberalism simultaneously accelerate environmental breakdown and pose significant barriers to an adequate response. The imperative for firms to expand turnover and profits drives the rise of material throughput and energy use in the economy. The exponential extension of markets to new domains has fuelled the commodification of nature, transforming all life into a potential source of profit through relentless geographic expansion and economic integration. Rather than maintaining and sustainably growing stocks of economic, social and environmental assets, the maximisation of income flows is central to current economic common sense, which prioritises short-term returns over long-term viability. It is a logic that drives the clearing of forests and the burning of fossil fuels for immediate gain, heedless of the self-defeating destruction that ensues. The extractive institutions and incentive structures at the heart of neoliberalism – concentrated shareholder ownership, over-mighty finance, the dominance of massive multinational corporations and their vast balance sheets, legal forms that grant capital owners exclusive economic coordination rights – have transformed our economies into engines of upward wealth extraction. In the heartlands of neoliberalism – Anglo-American capitalism – real wages have fallen behind productivity growth even as the 1 per cent have accumulated wealth on an extraordinary scale. Globally, it's estimated that fewer than thirty people have amassed more wealth than is owned by the entire poorest half of the population.[5] Wealth doesn't 'trickle down': it gushes upwards.

The deep inequality this generates has enabled wealthy individuals, corporations and industries to influence government policy by privately funding political campaigns, lobbying, and oiling the 'revolving door' between government and industry. This tangled web of material and political inequality allows vested interests to defend and extend their privileges, regardless of the wider environmental consequences. While wealthier countries may have cracked down on industrial smog and cleaned up rivers, they are still the largest emitters of greenhouse gases and, through the supply chains that feed their extravagant consumption, spur environmental breakdowns elsewhere, outsourcing their destructive impacts abroad.

Neoliberalism also assumes that humans predominantly act as rational, self-interested beings – the mythical *homo economicus*. Where they patently do not, neoliberal governance devotes considerable effort to creating and sustaining the competitive market subject, for example in the marketisation of higher education or healthcare. This erodes the potential for collective political action as the organising principle of society. Yet, though selfishness and self-interest are certainly part of human behaviour, we also have the capacity to be social, cooperative and adaptable, and to prioritise intrinsic over extrinsic values. Our extraordinary capacity for kinship, care and mutual endeavour are the very qualities that an effective response to breakdown demands.

But, as with the paradigms that preceded it, neoliberal common sense is fracturing, its contradictions and crises thawing the ideological permafrost that for decades claimed there was no alternative. The 2008 financial crash revealed the system's fragilities, including the co-dependency of private wealth on public protection. But it also demonstrated its adaptive resilience.[6] The subsequent return to business-as-usual, enabled by unprecedented monetary policy interventions, China's growing centrality in stabilising the world economy, and the political embrace of austerity, allowed neoliberalism's destructiveness to persist. Stark inequalities of wealth and power continue to consolidate into oligarchy. Finance remains over-mighty and systemically risky. The destabilisation of global natural systems accelerates. Even if it could rapidly steer us away from catastrophe, neoliberalism has created a society that is fundamentally bad for us – generating high levels of inequality, enormous power imbalances, poor mental health, social exclusion and extreme political polarisation, to name but a few of its worst effects.[7]

Covid-19 further exposed the brittle qualities of neoliberal governance and the limitations of financialised capitalism. The pandemic struck at the fundamental relationships of capitalism, turning off the engine of commodified labour, demobilising production, and subordinating markets to public health needs. In so doing it has revealed, once again, that the economy is not something prior to society, but an object constituted by law and politics, something public power can reorder towards a desired purpose.

And it has starkly exposed the limitations of neoliberal governance techniques. 'Nudges' and enforced marketisation cannot safely address the pandemic – in much the same way as they cannot defuse the wider structural crises of environmental breakdown and inequality. The crisis underscored our interdependence, and the case for an urgent reimagining. We can do better. In order to navigate the age of environmental breakdown – and build societies of economic and environmental justice – we must move beyond neoliberalism.

Thriving Not Surviving

Reimagining our future requires understanding that we stand at a critical juncture. Maintaining the stability of natural systems requires us to transform the political-economic paradigm. Permanent as the institutions of the present feel, we know deep change is possible: it has happened before. Dominant paradigms lose their legitimacy, often as a result of the failure of ruling ideas, classes or technological systems to adequately adapt to change or respond to crises.[8] Alternative ways of organising society emerge, capable of shaping the popular common sense, and erode then supplant the established order. At least in the UK and the US, this has happened twice in living memory. The consolidation of the post-war consensus – of the warfare and welfare state and regulated capitalism – replaced a discredited 1930s laissez-faire liberalism, aided by the new forms of solidarity and economic planning that emerged during the Second World War. The post-war settlement, which generated rising living standards and falling inequality in the West even as it exploited an unequal global economy, was in turn challenged and displaced by the neoliberal New Right. Seizing upon the contradictions and crises of social democracy in the 1970s, they mounted a successful counter-attack. The institutional transformation affected by neoliberalism remains fundamental to how our economies operate today.

Moving beyond neoliberalism requires learning from its success: if we want to secure deep, enduring change, we have to transform the foundational institutions that organise the economy and distribute power. Incremental tweaks of an economic model driving environmental breakdown guarantees a deepening planetary emergency. But this transformation must be faster and more fundamental than that which characterised previous paradigm shifts. First, we need to move to societies of genuine sustainability as rapidly as justly possible, in a matter of decades, if not earlier for wealthier nations, and over a period in which the destabilising impacts of environmental breakdown will grow. Second, however much neoliberalism has broken with the post-war consensus, there are nonetheless vital continuities with previous paradigms: the treatment of nature as a boundless resource from which to extract, a reliance on the unwaged work of social reproduction, the use of environmentally unsustainable technologies, the promotion and promise of high material consumption, and a racialised, gendered and hierarchical global economy geared towards Western interests. These continuities must end. 'Green Keynesianism' is an important step forward and carbon taxes have a role to play, but they will not be enough; maximising GDP growth as the primary goal of the economy is demonstrably inadequate, and the crisis of multilateralism requires new forms of international coordination. Systemic crisis demands a systemic response.

Third, deep economic change is upon us in any case. The scale of disruption prompted by Covid-19 means economies are already radically restructuring. Incrementalism is embracing defeat and ceding the ground to disaster capitalists. A collective act of unprecedented reconstruction is required, the reimagining of our societies to centre the needs of people and planet. Within little more than a decade, we must move beyond not only neoliberalism but the infrastructures, institutions and ways of life of extractive capitalism, ending enduring hierarchies, inequalities and injustices going back centuries. Our response to an age of environmental breakdown must be more than a set of discrete policies to decarbonise the economy; it must be a common project to transform society so as to put life first. If the environmental crisis is a political crisis, generated partly by the absence of constraining forms of democratic power over the economy and the inequalities that this has caused, our response must be political. *Contra* neoliberalism, we must democratise economic life to survive, decommodify to thrive, and build a commonwealth of solidarity and care in lieu of private power. This is not just about providing a counterweight to ethnonationalism and all its barbarity. It is about seizing our last chance to rediscover hope and cooperation from the wreckage of neoliberalism.

From Extractivism to Ecosocialism: The Way Ahead

What we are arguing for then is a deep and permanent reimagining of society in the face of existential crisis. This is ecosocialism, which for us means the collective effort to organise life through democratic power, with social control not private rule shaping the institutions and technical systems that govern and reproduce life, repurposing them towards the pursuit of collective and individual human flourishing that is inseparable from a thriving natural environment. This is a politics for genuine self-determination and emancipation. To that end, we seek to dismantle hierarchies of wealth, class, gender, race and power in society, replacing them with democratic relationships and powerful collectives. We recognise that 'nature' is political;[9] if the economic is inseparable from the environmental, if social and ecological relations are irreversibly entwined and co-produced, and if we inescapably intervene to restructure the Earth's natural systems, we should do so based on a politics that seeks to promote the thriving of life, both human and non-human.

That goal demands a different type of economy. One reoriented to giving social and environmental needs precedence over profit maximisation; to putting communal flourishing first, extending new models of democratic ownership and control, and realising deeper, more empowered democracies. Ecosocialism consequently stands for fuller, more sustainable forms of abundance and liberty than the partial freedoms, imbalances of wealth and power, and ecological crises generated by extractive capitalism.

Capitalism is a system for organising economic activity based on three foundational elements: private (and unequal) ownership of the means of production; production organised for the market to make profits and increase accumulation; and work is primarily undertaken by those who do not own the means of production. Within this architecture of power, those owning capital monopolise coordination rights within the economy.[10] By these means capitalism has generated vast, if unequal, wealth.[11] It has also driven extraordinary increases in productivity and material throughput, improving living standards for many, even as it generates poverty amid plenty. In doing so, it has helped create or strengthen freedoms, however limited, and has often coincided with the extension of formal political democracy. In many ways, capitalism – or neoliberalism, in its latest formation – is a spectacular mechanism for coordinating production and exchange, but one in which cruelty and injustice are ingrained.

Yet capitalism has also driven us to environmental breakdown. If we want richer, more universal forms of freedom and democratic power on a thriving planet we must go beyond the options that capitalism offers. As the sociologist Erik Olin Wright argued, 'through the functioning of its most basic processes, capitalism generates severe deficits of both freedom and democracy that it can never remedy. Capitalism has promoted the emergence of certain limited forms of freedom and democracy, but it imposes a low ceiling on their further realization.'[12] Ecosocialism consequently seeks to overcome six structural features endemic to extractive capitalism that inhibit collective flourishing and individual self-determination.

First, freedom must include the ability to say no. Yet most of us are required to sell our labour to live, our working days structured by employment relationships that are becoming increasingly authoritarian in nature, sharply circumscribing autonomy. Economic inequalities generated by 'private' action under capitalism undermine the capacity for meaningful individual and collective self-determination. Instead we need a social society, of equality, deep freedom and non-domination where people have the power to say no.

Second, many people are excluded from vital decisions, decisions that determine the quality and direction of their lives, communities and environments. Most fundamentally, private ownership of the means of production grants a privileged few the right to decide on where to invest or divest based solely on their self-interest. This privatises a vital power in society: the ability to order and determine our futures through control of investment. Given that existing patterns of investment are driving environmental breakdown, this is a striking deficit.

Third, waged labour can be organised around patterns of domination, hierarchy and control. The workplace and work are, for many, spaces with limited freedom of movement, voice and control, based on the entrenched inequality between labour and those who own capital. Constrictive hierarchies also shape non-waged labour and the distribution of freedom and power within – and beyond – the household economy.

Fourth, public authorities are pressured to bow to the needs of major businesses and investors, with a race-to-the-bottom culture limiting the scope for public action to shape our societies.

Fifth, deep economic inequality translates into sharp inequalities in political power, undermining the premise of political equality and limiting popular sovereignty.

Finally, the most dangerous limitation imposed by extractive capitalism is that its cumulative impacts are driving climate crisis and the accelerating collapse of natural systems. There is no freedom on a dead planet.

These limits do not mean capitalism cannot be contested or improved, with new rights and freedoms secured. As Olin Wright argued, the boundary between democratically determined public rules and the realm of private freedom is shaped by politics. Indeed, social democratic and socialist movements have won vital extensions in freedom, from collective bargaining to the regulation of corporate power, to the security a decent welfare state can provide. But limits on freedom and democratic power are almost impossible to fully eliminate, due to the fundamental arrangements of capitalism. There is a duality to capitalism. It at once expands our ability to consume (for those with the means), creating a world of seemingly bound-less choices. Yet it also limits the horizons of freedom through its concentration of wealth and power and the snares of debt and hierarchical wage relationships it generates. This contradiction – both enabling and contracting our freedom – is arguably most pronounced under neoliberalism.

The core of neoliberalism is a multi-pronged strategy to insulate capitalism from democracy, transforming the economy into an object beyond the scope of political inter-vention. We must realise its opposite: the re-embedding of the economy within natural systems, a recognition of how institutions can be changed by politics, and a belief that deepening the capacity for individual and collective self-determination is a precondition for a just, sustainable future. In this, we seek the same goals as democratic social-ism, whose roots ecosocialism shares: 'an obstinate will to erode by inches the conditions which produce avoidable suffering, oppression, hunger, wars, racial and national hatred, insatiable greed and vindictive envy'.[13] Yet, given the time-critical nature of our crises, we must seek to quicken the pace of political life, aiming for a deeper, more rapid reordering in our economic and social institutions and relations. Nor does ecosocialism share the productivist impulse of some twentieth-century socialist traditions, which often matched capitalism's tendency to treat 'nature' as an external, endless 'other' and relied on an overly centralised state as the main institution for change. And it seeks to move beyond the limits of certain forms of labourist culture: too often male-dominated, hierarchical and privileging certain forms of work.

Our vision of the future is founded on the democratic reimagining of the four fundamental institutions of production and exchange in our societies – the commons, state, household economy and market – and of how they might interact to provide the conditions for thriving. We seek an expanded commons, stewarding the resources and infrastructures – natural, social and digital – we all need to live well. Instead of austerity and a repressive state, we seek a repurposed, democratised state that invests in a liveable future, actively restructuring economic activity towards social and environmental justice through green industrial strategies, expanding the decommodification of society through the provision of universal basic services and prosecuting an agenda for twenty-first-century public ownership. We fight for a household economy where all forms of work are properly valued, in which burdens reduced and redistributed, and illegitimate hierarchies challenged. Against the rule of capital, it seeks a democratised marketplace, with finance under social control, enterprise reimagined as a generative institution, and underpinned by a pluralistic ecosystem of common ownership and control.

This may seem a radical reimagining, but in an era of environmental breakdown and structural inequality, transformative action is the safest and fairest path forward. A dogged defence of the status quo will guarantee the acceleration of environmental breakdown; a willingness to only tweak a model driving us deeper into crisis is the truly extreme position.

We need to learn and draw energy from a wide coalition – intersectional, anti-oppression social movements that centre class, gender and race analysis in the fight for collective and individual liberation, radical liberal traditions currently muted by centrism's self-defeating timidity, and social democrats committed to building an entrepreneurial, ambitious state. A key argument unites these movements: what is needed is not charity, but instead the assertion of economic and environmental justice, redistributing to the many what is unfairly taken by the few, organising the world more justly, efficiently and effectively.

The movement beyond neoliberalism – as a first step – requires accepting that there will be no final confrontation with contemporary extractive capitalism, no binary moment where it is overthrown. Nor is there a systemic way beyond it via the politics of escape, whether a folk politics of eco-primitivism or technological moonshots. Instead, we must pursue two approaches in tandem to build societies of environmental and social justice: at once creating or expanding existing post-capitalist institutional forms, while taming capitalism through reform of its foundational institutions, from transformed public investment to the re-embedding of finance in the real economy.[14] We therefore require confident use of the tools neoliberalism has long sought to constrain: collective action, ambitious public investment, strengthened labour power, democratic planning and democratised workplaces, the extension of the democratic public realm and common ownership.

This is not just about providing an appealing counter to rampant ethnonationalism, stepping up where timid centrism fails. It is about recovering a future of radical hope from a planet on fire. The next five chapters explore what that could look like, imagining new ways to cooperate, invest, own and live in a world of flourishing and meaning.

4

AFTER EMPIRE

Will we succumb to chaos, division and inequality? Or will we right the wrongs of the past and move forward together, for the good of all?

We are at breaking point. But we know which side of history we are on.

Antonio Guterres, UN secretary-general

We've been here before, some say; environmental concerns are not new. Two truths can be discerned from this timeless story, it is asserted. First, the fear of catastrophe is ever present, always influential, and often unfounded. Second, human ingenuity and cooperation eventually win the day. It is a reassuring narrative, and rarely does a week go by without an article or a flurry of tweets urging us simply to relax, trusting to humanity's past form in overcoming environmental odds. Often, these arguments point to the Montreal Protocol as proof of such a record. The Protocol was the culmination of an acute period of environmental concern. Over the 1970s and '80s, scientists discovered that chlorofluorocarbons (CFCs, chemicals used in air conditioners and aerosol sprays) were eating away at the ozone layer. This layer acts as a planetary shield, protecting humans, animals and plants from damaging ultraviolet radiation that would otherwise stream through the atmosphere. CFCs, the scientists found, were reducing the concentration of ozone in the atmosphere, thinning the protective layer and, in places, creating 'holes' through which more radiation could reach the Earth. This was big news and people were right to fear the potentially catastrophic global consequences of ozone depletion, from cancer to collapsing crop yields. Recognising the great benefits of sustainability – and the greater costs of inaction – the United States led the way in securing a global agreement that applied the precautionary principle. In 1987, less than two decades after scientists sounded the alarm, the Montreal Protocol was signed, ordaining the phase-out of ozone-depleting substances. Today, the southern ozone hole is at its smallest for decades.[1] The Protocol is seen as a model of cooperation under conditions of environmental peril, one of the most successful international treaties in history, if not the most successful. This remains true, and is therefore a tragedy.

As the Montreal Protocol shows, cooperation is one of the essential components needed to avoid environmental disaster. Governments must negotiate treaties, navigating the complex minefield of competing interests. Companies must respect laws and change their plans to meet the environmental imperative. Each individual must play their part in the collective effort. Those trying to allay our fears of environmental catastrophe rightly point to previous examples of cooperation, such as the Montreal Protocol, as models for how we can avoid the worst. But this view is founded on a critical failure to recognise how cooperation in the age of environmental breakdown is both uniquely difficult and uniquely threatened.

For a start, the nature of the problem is different to anything we have faced before. Environmental breakdown is pervasive, occurring on an unprecedented scale, encompassing local biodiversity loss and the destabilisation of global systems. It is accelerating and some natural systems may have already been pushed beyond points of no return. Environmental destruction is intertwined throughout the structures of our societies and economies. As a result, those countries who must act first and fastest are often the greatest beneficiaries of these structures. This creates a collective action problem, which tends to corrode cooperation between countries and communities. If rich polluters who caused the problem and are best placed to act will not, then why should others, and if they did, what difference could they make? This problem cuts from the international level down to the local, pitting us against each other. The result is a global standoff, which has now persisted for three decades.

Even with ambitious net zero announcements, a stalemate persists. Ours is a world vastly different to that which bequeathed the Montreal Protocol. It is more unequal, power has been fragmented and privatised, and we are far less trusting. A focus on individual responsibility belies the power of the collective, collapsing action down to consumer choice, fixing attention on paper straws and 'woke brands' over the structural drivers of breakdown. The ongoing effects of the 2007–08 Global Financial Crisis have translated into social and political destabilisation, jolting geopolitical vectors away from international cooperation and championing leaders who eschew collective action. The parlous record of international cooperation at times over the course of the coronavirus pandemic was concerning. These developments have come at precisely the wrong time. Ours should be a golden age of collective action, the heir to the triumph of Montreal. Instead, the room for cooperation is closing as the carbon budget runs down and the seas choke with plastic. Each year of inaction means that the actions needed to avoid the worst become more extreme. Emissions climbed in 2019, so the reductions up to 2030 would have to average 7.6 per cent across the world, and around 15 per cent a year if we don't assume a vast clean-up effort will be undertaken in the future. As emissions rise, the rate of decarbonisation required could get steadily more impracticable for the wealthy and impossible for the poor.

All this will occur over a period of mounting destabilisation wrought by environmental breakdown, reducing the incentive to work together and favouring the cause of ethnonationalism. In response, we need a new type of cooperation fit for our new age of environmental break-down, an age far from the relatively stable circumstances of the late twentieth century. It must be one that unlocks the potential of communities and countries across the world and binds them together in a common purpose.

Communities of Resourcefulness

Acting to slow environmental breakdown is often seen as the preserve of cooperation at the international level. But without the engagement and concerted action of localised communities – villages, towns and cities – grand, global designs will remain just that. Many of these communities feel disenfranchised today. Promises to 'take back control' and to make countries 'great again' chime with the lived experience of millions of people betrayed by the promises of neoliberalism. The neoliberal management of techno-logical change and globalisation has largely served the interests of a small elite, eliminating jobs and eroding the basis upon which many communities were founded. Misguided economic policies have frequently exacerbated these trends, running down essential services while allow-ing big business to extract a growing share of the wealth created locally. To people living outside of narrow elites in London or New York, control is something others have. The rise of ethnonationalism is partly fed by ignorance, liberal indifference to the erosion of agency felt by so many.

This is one of the central reasons why communities must be re-enfranchised if we are to have a chance of confronting environmental breakdown. Communities that work and care together, with greater agency over their own development, are better able to beat back the regressive political forces emboldened by the growing destabilisation. But this is about something more than 'resilience'. The injunction to become more resilient is usually voiced by custodians of the current system, those with a vested interest in perpetuating the very factors that make the system brittle. Their favoured conception of resilience is imposed top-down, a rigid appeal to weather the storms of a system that punishes those least responsible and most at risk, protecting the inequalities that perpetuate the status quo. Instead, the resourcefulness of communities must be unlocked so they may realise alternatives to the system; only then can they become more resilient to the destabilisation to come.[2] Moving beyond neoliberalism at the local level can prompt a more effective, distributed effort to slow environmental breakdown. This can create the conditions for an 'all-society' response to the emergency, one in which we mobilise the full resources of society to tackle environmental breakdown. Such a response has only been seen in wartime and nearly came about in response to the coronavirus. It has not been tried yet and, as the environmental crisis has reached its terminal phase, is now essential.

Local people can identify local environmental problems and their contribution to global trends, collectively formulating local responses that are realised through harnessing local agency. Local, sustainable food production shortens supply chains, reducing the environmental impact of consumption, while protecting communities against price shocks. Community energy production keeps local money from flowing offshore through the creative accounting of energy giants. Nature restoration projects boost community engagement, with people reaping the benefits of a more meaningful relationship with the environment. The mutual aid societies and myriad ad hoc networks of community support that grew up in the wake of the coronavirus pandemic are heartening proof of local agency and its effectiveness.

Fundamentally, resourceful communities come about when local areas are given more oversight, democratic input or outright control over economic and social institutions, from energy through education to health. This logic flies in the face of neoliberal policies, which have privatised and centralised the running of these institutions. Increasingly, local communities are fighting back against the failure of this approach. A leading example in the UK is the city of Preston in the county of Lancashire, which has developed a model that leverages the power of public institutions and the capabilities of local people.

It started with local public bodies – the city and county councils, educational institutions and housing associations – looking at how and with whom their money was spent ('procurement' in the jargon). Out of a total spend of £750 million, only 5 per cent went on organisations in Preston and 39 per cent was spent on those in Lancashire.[3] This was partly the result of rules putting price above all other considerations, with contracts going to big businesses from far beyond the county, hollowing out the local economy. Competition crowded out cooperation and social outcomes came second to 'value for money'. This was a false economy, a wasted opportunity for keeping wealth in the hands of local people and building a stronger local society. So, the public bodies of Preston began to change their procurement rules and divert spending towards organisations and projects in the local area that met a wider range of criteria. These included whether they provided secure, well-paid jobs, or would reinvest in local buildings and land to maximise social benefits, rather than profits. Here the public bodies were using one of their greatest powers − spending − to improve the well-being of local people. The bodies also 'insourced' public services, reversing the trend for outsourcing services to large multinational corporations, and, in doing so, offered decent wages and apprenticeship schemes. They also encouraged the creation of cooperatives and other ways of increasing local, shared ownership of the economy, and invested their pension funds in socially useful projects. This strategy has transformed Preston. Between 2012 and 2017, an extra £200 million was invested back into the local economy, increasing locally retained spending in Preston and Lancashire to over 18 per cent and 79 per cent respectively.[4] Thousands of people have been brought onto the Real Living Wage, education has improved, in-work poverty has reduced, Preston has become less deprived, and, in 2018, it was named the 'Most Improved

City in the UK'. All this was achieved under austerity imposed from Westminster.

Preston joins similar efforts across the UK and around the world, some of which served as the original inspiration for Preston itself. The US city of Cleveland has long been applying a similar model, in which public institutions purchase goods and services from worker cooperatives who hire members of the local community.[5] These institutions also make patient investments in non-profit companies which, in turn, partner with the local government to finance the worker cooperatives. Such examples offer a different, post-neoliberal model for how the economy can be designed and how and for whom it works, known as 'community wealth building'. At its heart, community wealth building provides the means to rescue cooperation from the ravages of neoliberalism, offering a chance for local people to actively intervene in their local economy to realise goals they set, not just managing it and leaving its development (or lack thereof) to the whims of large companies. This is what makes community wealth building so notable. By applying its models, people don't have to wait around. Instead, they can transcend the failure of national governments and corporations and build a new society in the process.

To date, the focus of community wealth building has largely been on economic justice, ensuring wealth remains in local communities and that its people share equally from the opportunities that arise. But community wealth building is tailor-made to drive rapid sustainability from the bottom up, giving everyone a role to play in a great global effort to respond to environmental breakdown. Locally decided rules can ensure that investments are channelled towards clean development, and that public money is used to support organisations that heal environmental destruction while providing better public services. Green businesses and products and services that promote wellbeing and community over endless material consumption can be encouraged by currencies that can only be spent in local areas, or through neighbourhood banks that prioritise investment in these ventures. Key utilities can return to shared ownership, allowing water, energy and transport to be supplied in a way that prioritises people and planet over profit, as is the case with many municipal energy and transport companies around the world.

A Coordinating State

Local communities can only do so much in the face of global forces. Clearly, we cannot respond to environmental breakdown and assemble an all-society response without realising the full potential of the institutions of government. Governments are endowed with powers that are uniquely placed to resolve collective action problems and achieve goals beyond the shorter-term focus of market behaviour. Moreover, as the coronavirus pandemic has so starkly reminded us, governments are able to collect and deploy vast resources with a view to long-term needs, such as establishing public health systems or developing clean technology in which they can invest, reducing the risks of doing so for others. Governments can identify overall goals for societies and coordinate the investments and action required to meet them, as happened during the two world wars or in reaching the Moon. Above all, governments can be motivated by a democratic impulse to seek 'non-market outcomes', such as promoting political or gender equality and protecting minorities. In turn, governments deserve credit for positive responses to environmental breakdown that would have been impossible without them. Legislation, such as clean air laws or acts to reduce greenhouse gas emissions, has placed restrictions on the environmental destruction caused by economies. Unprecedented resources have been ploughed into researching, developing and deploying clean technologies. Constitutional rights to the benefit of a healthy environment have been introduced around the world. We would be nowhere without the UN climate process, whatever its failures.

That said, the response of governments has been glaringly inadequate. This is partly a consequence of the neoliberal conception of the state that has come to dominate over the past forty years. This dictates that the role of government should be limited to the correction of 'market failure', based on a strong, market-enforcing state. It will not countenance a government ready to invest and intervene on the scale likely needed to reduce emissions by 7.6 per cent per year and stem biodiversity loss. The corollary of this attitude is a reduction in state capacity, a self-fulfilling prophecy of austerity, outsourcing and privatisation. The neoliberal model of government is partly responsible for environmental breakdown, crippling state capacity at precisely the wrong time, allowing short-term market interests to sway the calculus of government decision-making. Where state power remains, it is centralised, stifling the development of resourceful communities.

In looking for a path beyond the neoliberal state, we find one national institution that has always embodied the alternative to neoliberalism – the National Health Service (NHS). The NHS is Britain's publicly owned and provided system of healthcare, free at the point of use and funded through general taxation. Debates about the NHS (rightly) focus on funding and its stealthy privatisation. But this has obscured important dimensions of the NHS's work: its impact on society goes beyond treating people, and, further, demonstrates how a different model of government could accelerate action against environmental breakdown while redressing its harms.

As of 2018, the NHS spent £27 billion on goods and services in England each year, was one of the largest employers in the world, with 1.5 million staff (around 5 per cent of total people in paid work), and occupied over 8,000 sites across 6,500 hectares in England alone.[6] The sheer size of its operations means that the NHS has an influence on the world beyond making sick people better. This can be good or bad. For example, the NHS has firm views on what constitutes value for money when it comes to drug prices, an assessment used by both the centralised NHS drug procurement system and many medical services around the world, which pushes down the global price of drugs.[7] On the other hand, the NHS is responsible for around 3.5 per cent of all road traffic in England at any one time, resulting in greenhouse gas emissions and air pollution that cause ill health, shorten lives and impose an estimated economic cost of £345 million.[8] Either way the NHS can use its enormous power to improve social, economic and environmental outcomes in Britain above and beyond the clinical sphere.

Crucially, this can be done in a way that encourages the flourishing of resourceful communities, moving beyond the centralisation of the neoliberal era and the post-war consensus before it. This is, in fact, already happening. Several hospitals and health centres are adopting the community-wealth-building approach to behave as an 'anchor institution', leveraging their important role in local areas as purchasers of goods and services, landowners and employers. This has led some to install renewable energy on hospital sites funded by community investment. This pays out a return to local people, and, in some cases, profits go into a fund to support those in fuel poverty. In turn, wealth is created and retained locally, greenhouse gas emissions are reduced, energy costs fall, and health improves as fuel poverty is reduced. Meanwhile, other NHS bodies have implemented schemes to support staff and patients in using shared transport and in cycling and walking, gaining fitness from less air pollution and more physical activity while lowering the transport bill. Other examples include procuring some of the 300 million meals the NHS serves each year from sustainable, local food producers, or recycling waste heat to local people, reducing bills and saving energy. The list of projects goes on and, in each case, shows that more sustainable healthcare equals more efficient and effective healthcare. Overall, local action as well as system-wide changes, including changing how central NHS bodies behave, has helped the NHS reduce its environmental impact to a remarkable degree. Between 2007 and 2017, NHS emissions fell by nearly a fifth, an amount the same as the yearly emissions of Cyprus. It managed to do this while growing in size by over a quarter, according to some indicators of the NHS's clinical activity.[9] Between 2010 and 2017, it reduced its water use by over a fifth, the equivalent of 243,000 Olympic swimming pools, and around 85 per cent of waste no longer goes to landfill.

The NHS has a long way to go before functioning within truly safe environmental limits. But, in many respects, it leads the world in understanding its environmental impact and acting to reduce it. In the process it shows us the indispensability of the state in responding to environmental breakdown and doing so in a fair, better-prepared way. A singular healthcare system allows for more rapid change than do fragmented private systems as in the US, while also ensuring that healthcare remains a right and not a privilege, provided on the basis of need, not ability to pay. In this way, the national structures of the NHS could provide equitable protection from the

enormous threats to health posed by environmental breakdown, in much the same way as they did in response to the
coronavirus pandemic, marshalling extraordinary
resources to build entire hospitals in a matter of days. It is
not just the Chinese state that can do so. And these structures can also enable the NHS to act as a powerful nation-
wide driver of sustainability while, along the way, reaping
the enormous benefits to health from adopting more
sustainable diets and ways of travelling, as part of lower-
impact lifestyles in general.

Such a vision for the NHS provides us with an alternative
model for the state as a whole. It is a model that underpins
the conception of a Green New Deal – the plan for enormous state-directed investment to combat the climate and
other environmental crises, while rectifying structural injustices and kick-starting stagnant economies. The necessity of
and opportunity for a green and equitable recovery in the
wake of the coronavirus and the breaking of the myth that
governments cannot borrow and finance to invest on an
extraordinary scale to achieve pressing goals must kill off
any excuses for not implementing a Green New Deal. But
even then we are very far from the adoption of a post-
neoliberal model of the state. As with the NHS, the sale of
public assets and the outsourcing of its functions have steadily eroded the power of the state. So have major budget cuts
and 'reforms' that have led to public services being run like
businesses, creating a wasteful bureaucracy that enforces the
neoliberal mantra that competition and not cooperation is
the route to success. Moreover, genuine democratic

oversight eludes our creaking nineteenth-century political systems, which restrict the voice of voters and favour the influencing agenda of vested interests. In 1948, Britain democratised good health. With the NHS, generations of struggle had gifted people the freedom from fear. Generations later, greater democratic oversight over economy and society is a prerequisite for winning freedom from the fear and fire of environmental breakdown.

After Empire

As the coronavirus pandemic has so brutally reminded us, no country is an island. Just as the fate of each person is intertwined with those around them, the lives of nations are determined by the great global systems binding their actions to those of others. Foremost among them is the modern globalised economy, with its treaties, financial flows and multilateral institutions presiding over a web of interconnection that links banks in London to subsistence farmers in Africa. Over the last four decades, a neoliberal model of economic globalisation has evolved out of the end of formal empire, protecting and extending the inequalities and exploitation of the imperial era and accelerating its destruction of the environment.

Iniquitous financial arrangements permeate this system. Since the end of colonialism, loans, often on exploitative terms, were foisted on countries in the Global South. High debt repayments impair countries' ability to pay for healthcare, education and environmental protection and make them vulnerable to continued exploitation by other governments and private companies. Over thirty countries are now in a debt crisis in which repayments have become unmanageable, including countries which have, like Mozambique, also suffered cyclones and other extreme environmental shocks. It was estimated in 2015 that Sub-Saharan African countries received over $160 billion in loans, aid and investment from abroad each year, but lost $203 billion from tax avoidance, debt repayments, extraction of resources, transnational corporations taking out profits, and the growing costs of climate and other environmental emergencies to which the continent had contributed little – a net loss of $43 billion.[10]

Wealthy nations play a key role in this state of affairs. For example, Britain is responsible for legal structures governing the lending boom, with as many as 90 per cent of the publicly traded loans to governments in Africa made under English law.[11] In addition, tax havens, many of them in the outposts of what remains of the British Empire, are used by corporations across the world to avoid paying taxes. Upwards of $7.5 trillion of wealth resides in tax havens, instead of being invested in projects to build a more sustainable, just and prepared world, and governments are losing over $190 billion a year in revenues.[12] These countries have been plunged into an unprecedented economic crisis by the coronavirus pandemic, significantly worsening the situation.[13]

Meanwhile, multilateral institutions, such as the International Monetary Fund (IMF) and the World Bank, encourage corporate exploitation of countries across the Global South through 'structural adjustment' programmes that demand austerity and privatisation in return for financial assistance in times of crisis. Privatisation policies hand over key assets to private companies, a policy which has been described by the UN as 'systematically eliminating human rights protections and further marginalising those living in poverty'.[14] The governance of the World Bank and IMF is dominated by wealthy nations, including through the 'gentleman's agreement' that the organisations' heads are always American and European respectively.

These power imbalances are further entrenched by treaties that put the rule of the corporation above the rule of law. A prime example is the Energy Charter Treaty (ECT), an international agreement which allows corporations to sue countries for changes in energy policy, with the resultant cases – or 'investor-state dispute settlements' (ISDS) – taking place in private courts and presided over by private lawyers away from public scrutiny. The ECT has proven lucrative for energy companies, who readily sue governments for billions of dollars, and has been used to halt or roll back environmental regulation.[15] In all, it has protected fossil fuel investments totalling over double the greenhouse gas emissions the EU is allowed to produce if temperature rise is to be kept under 1.5°C.[16]

Overall, these (and other) structures have created a global economic system that enthrones those governments and corporations that profit from environmental breakdown, legitimising their actions. In doing so, it has accelerated environmental breakdown and entrenched inequality between those who cause the problem and those who experience the worst effects. The devastating impact of the coronavirus pandemic and its economic consequences for the Global South and many emerging economies is a strong function of these arrangements.

As destabilisation grows, people and countries find themselves pitted against one another at precisely the wrong moment. Yet the unjust structures of the global economy and their impact are often missing from analyses of the current crisis. The failure to spot the straitjacket of neoliberal globalisation constricting our response to environmental breakdown is the greatest expression of the trap of the ecocide narrative of Easter Island. It is hard to see how the standoff we observe at UN environmental conferences will be resolved without modifying the system. Meanwhile, its failures benefit those who seek to tear apart international cooperation. Denialist Conservatives and Eco-ethnonationalists are already fracturing institutions of global cooperation, so as to further entrench unfair trading and financial regimes.

In response, we need a positive-sum internationalism, under which the primary aim of cooperation between countries should be to rapidly reform global economic structures. This will have to be done while managing the growing destabilisation of environmental breakdown, maintaining cooperation against the lure of isolationism.

The first step is telling an honest story that links the past to the present. To some extent, this is already happening;

the UN processes of international environmental negotiation enshrine the principle of 'common but differentiated
responsibilities'. But there is no compelling story here: the
histories and realities of injustice are obscured behind
technocratic jargon and closed-door negotiations. Under
conditions of environmental breakdown, successful international cooperation must be founded on an accessible and
persuasive story that no longer excludes vast swathes of
humanity, leaving them to suffer and stoking their resentment. This is particularly important when considering that
the recent economic development of poorer nations has
blurred the categories, giving wealthier nations licence to
demand more of China, India and other recent emitters.
Countless meetings have ended with representatives of
Global South nations aghast at calls on them to end coal
power or deforestation, coming from nations that still
enjoy the benefits of having done precisely the opposite.
These demands are not wrong, but they are often made
without candour on the part of wealthy nations. Truth and
honesty between nations has always been hard. But our
future depends on it more than ever. Countries that grew
rich through the unsustainable use of resources on a scale
that has driven environmental breakdown must take more
definitive steps in a new direction, telling an explicit story
about the links between their own past actions and environmental breakdown, and how this history contributed to
present injustices. This necessarily means talking about
imperialism. Ten generations on from the advent of imperialism, its terminal phase sees the barbarity of the gunboat

and the slaver succeeded by that of the Category 5 cyclone and the Frontex patrol drone.

The UK, for example, is the fifth-largest contributor to the total greenhouse gas emissions released since the onset of industrialisation, and remains a major emitter. Its status as the world's fifth-wealthiest economy was made possible by these emissions. Moreover, Britain's contribution is also a function of its leading role in creating and entrenching a model of unsustainable environmental exploitation. This story should be backed up by action founded on explicit recognition of cumulative responsibility for the problem and the contemporary capability of nations to act. Facing its past necessarily requires Britain to take a fair share of responsibility for what happens next.

In the case of climate breakdown, this means playing a part in reducing the remaining global greenhouse gas emissions, besides simply reducing its own emissions. Having contributed more to the mess, Britain is that much more responsible for cleaning it up; it has accumulated a climate debt greater than its current carbon largesse. This can be done by simultaneously reducing domestic emissions to net zero or net negative in advance of 2050, with the great majority of reductions occurring this decade, in effect giving poorer nations more greenhouse gas emissions to play with, relieving some of the increasingly unmanageable burden placed on these countries. One estimate puts Britain's fair share of global greenhouse gas emissions at equivalent to around 200 per cent of domestic emissions below 1990 levels by 2030.[17] This would require a significantly higher ambition for 2030 than currently planned. A similar principle could be extended to include other measures of environmental destruction, including how consumption in Britain drives destruction of land and wildlife abroad through globalised supply chains.

Taking a fair share of cleaning up the mess also means helping those who need it the most. There are a range of ways Britain and other wealthy nations can do this. One is providing financial support, as many already do. Yet these contributions aren't based upon a nation's fair share. To do so, wealthy nations would have to raise their donations to international funds to support less wealthy countries at least ten-fold, to £20 billion up to 2030 in the case of Britain.[18] In a destabilising world, we can no longer afford to let wealthy nations disobey the rules of international law and sustainability and human rights commitments, such as the Sustainable Development Goals. Nor can we afford for them to be extractive bullies. Instead, they should act as supportive partners, using the safety and power that privilege brings, with the humility that their historical behaviour demands. This means eliminating tax havens and negotiating trade agreements that penalise environmental destruction and favour human rights. Countries that enjoy power in multilateral institutions should apply conditions to their input to the World Bank, IMF and others, to ensure these organisations become more democratic and escalate action to stem breakdown while promoting human rights and local capabilities. When greening at home, they must be aware of the impact of sustainability abroad, of the cobalt mines feeding electric vehicle supply chains or the farmland given over to homogeneous bioenergy crops. Sustainability in one nation cannot come at the cost of other communities and countries.

None of this can be achieved without opening up the arena in which international decisions are made. This means building alliances for justice that support the voices of global trades unions, civil society organisations and those representing communities on the front line of environmental breakdown – not those that continue to privilege wealthy interests in opposition to sustainability and justice. It is imperative that a wider constituency of people have a genuine say over the construction of the global economy. Without them it is hard to see who will, in the short time that is left, champion binding global agreements to collectively manage the enormous risks of an environmentally destabilising world, which the coronavirus pandemic foreshadows, and enshrine human rights over the rapacity of transnational corporations.

And there's another reason why this is needed. Communities in the Global South have been at the forefront of environmental breakdown for generations, and if we sail past 1.5°C and 2°C, a world of gross power imbalance will condemn millions more to death, effectively sacrificing them to fuel the high-consumption lifestyle of the wealthy for a couple more decades, until even they cannot escape the spiralling catastrophe. Even before this, millions could be forcibly displaced, fuelling the narratives of ethnonationalists who seek to justify the militarisation

of borders. It is therefore imperative that international
protection is given to those on the move in an environmentally destabilising world, that financial support be given to
those states receiving the displaced, and that trans-location
agreements be reached with the most severely affected
places, such as the Pacific Islands. But most of all, there
must be a shift in the narrative around migration. Most
people want to stay home, with migration a desperate, final
form of adaptation. We must create a larger 'us', a global
humanity facing up to a global predicament. While the
problem may have been caused by a few, we all live in a
world that suffers from it. This is why those voices marginalised for far too long must now soar. Those for whom
environmental breakdown is already an existential threat
must become world leaders, their struggle for justice and
survival acting to remind everyone of how we got here —
and how we get out of it.

5
FINANCING PLENTY

We cannot rule out catastrophic outcomes where human life as we know it is threatened.

David Mackie and Jessica Murray

Report on the Economic Risks of Climate Change,
February 2020

On 31 January 2020, Jim Cramer, host of *Mad Money* and doyen of American finance television, launched into a monologue that ended with him reading the last rites of Big Oil: 'I'm done with fossil fuels . . . they're just done.' Zeroing in on a sector in crisis, the clip quickly went viral. Beset by global overproduction, rising indebtedness, geopolitical tensions that had erupted into a price war between Russia and Saudia Arabia, and the rise of renewable energy and electric vehicles, the sector was losing value fast. Yet the implosion was only just beginning.

Three months later, something unprecedented happened: the price of West Texas Intermediate, the benchmark standard for US oil, went negative for the first time in history, at one point slumping to negative $37 per barrel. Traders were willing to pay people to take the oil off their hands, the barrels worth more than the content. Oil, the apex resource of fossil fuel capitalism – a fuel source whose cheapness and abundance had transformed calculations of finitude, restraint and the possibilities of long-term growth in the twentieth century; whose carbon energy had created new assemblages of political power, techno-institutional forms, and geopolitical configurations;[1] and whose extraction and burning is driving ecological collapse – appeared temporarily worthless. The death knell was clanging.

The reality was more complicated than headlines suggested, reflecting the volatility of the futures market and short-term storage costs; oil had not become valueless overnight. But the direction of travel appeared clear: the unprecedented collapse in global demand triggered by coronavirus had further exposed the deep structural weaknesses of the oil and gas sector, driving a wedge between the hitherto entwined leviathans of the extractive economy: the fossil fuel industry and the financial sector. Extraordinary as this moment was – and harbinger of deeper rupture that it may prove to be – the negative oil price shock was only one wild gyration in a financial event of world-historic proportions. As economies went one by one into enforced hibernation in response to Covid-19, the dollar-based global financial system, already under strain, experienced a profound shock, one that is likely to permanently and radically reconfigure the nature of globalisation. As the historian Adam Tooze reported:

> In the third week of March, while most of our minds were fixed on surging coronavirus death rates and the apocalyptic scenes in hospital wards, global financial markets came as close to a collapse as they have since September 2008. The price of shares in the world's major corporations plunged. The value of the dollar surged against every currency in the world, squeezing debtors everywhere from Indonesia to Mexico. Trillion-dollar markets for government debt, the basic foundation of the financial system, lurched up and down in terror-stricken cycles.[2]

This shock prompted an extraordinary policy response.[3] The world's central banks and treasuries deployed a rolling series of interventions that dwarfed in speed, scale and ambition their actions following the Global Financial Crisis. The Federal Reserve led the way; an actor of imperial sway operating from the heart of an increasingly dysfunctional empire. Swap lines to provide dollar liquidity to partner central banks were reopened, a critical tool of triage in 2008, now expanded to previously unseen levels. Central banks flooded credit and financial markets with liquidity and purchased government debt on a vast scale. Interest rates, already historically low, were driven down further across the world. Backstopped by central bank action, treasuries committed to unprecedented levels of economic intervention. By 23 March 2020, just twelve days after the outbreak was officially declared a pandemic, the Fed announced quantitative easing would go on indefinitely to support the stricken financial system. Within less than a fortnight, as the central mechanisms of global financial capitalism seized up, the boundaries of the possible had seemingly been abolished.

Third Time Lucky

Twice now, in little over a decade, public power has been mobilised to rescue a dysfunctional and extractive financial system. Each time, extraordinary fiscal and monetary firepower has been deployed and institutions radically re-engineered. It has been a deeply political project, with state power used to protect private wealth. The effects have dramatically reshaped the contours of the global economy and reconfigured geopolitics. Yet the scale and range of intervention each time has also provided an unexpected source of hope: it has underscored the plasticity of financial institutions and their political ordering, money's status as a contestable social relation, and the ability for politics to mobilise and direct the power of finance. What's more, the scale and speed of mobilisation has confirmed that the fundamental challenge of rapid transition is not primarily a question of resources or financial capacity, but rather a question of politics and of political will. The financial system is a vital utility, a public good, captured by private interest. If democracies can wrest back control over financial power, we can deliver all manner of transformative change that can breathe life back into our future.

If finance's power has been used to stabilise extractive capitalism in the past, we now need to harness this directing force to steer our economy towards comprehensive decarbonisation and the repair of nature, while building a society where all life can thrive. The largest mobilisation of resources in peacetime history will be required to build the infrastructure and institutions of ecological plenty from the ruins of fossil fuel civilisation. This cannot happen relying on a financial system whose current design and operation are driving breakdown. Instead, we need to socialise investment to deliver a rapid and just transition.

The democratisation of finance, bringing it under social control, must be at the heart of a Green New Deal: a transformative expansion in the scale and ambition of public investment to drive decarbonisation and repair nature, aided by a new network of mission-oriented public banks, to bring to life the green jobs and sectors of the future; the extraordinary planning power of central banking used to steer financial activity towards sustainability; and the power of private finance tamed, its potential refocused on serving social and environmental needs. Overarching this, a just architecture for global financial governance is needed, deploying policy tools the 'Washington Consensus' long forbade. Our future depends on nothing less. Under social direction, a democratised financial system can help resuscitate our world.

Morbid Symptoms

The worst crisis in capitalism's peacetime history exploded into a financial system already riven with weaknesses and contradictions. Many of its dysfunctions remained from the Global Financial Crisis of 2008: high levels of global corporate and private debt, a vulnerable Italian banking sector threatening the Eurozone's stability, historic and volatile asset bubbles, a banking system channelling too much credit into financial assets and real estate over productive investment, and a monetary system already at the limits of conventional firepower. Low interest rates and quantitative easing – another legacy of the Global Financial Crisis – had kept the wheels of capital accumulation turning in the 2010s. But though a flood of cheap money had maintained the economy on life support, the main beneficiaries were rentier interests: rising asset prices, an explosion in dividends and share buybacks, and a surge in industry consolidation that benefited wealthy investors, asset-holders and executives.

Perhaps most critically, the design and operation of the financial system was and remains a critical enabler driving environmental collapse. One of finance's key functions is the allocatin of capital among a range of possible uses. The institutions principally directing resources between competing demands today are private, profit-maximising banks and other financial institutions, including asset managers, private equity, venture capital and a range of organisations in the 'shadow banking' sector. They decide who receives finance and on what terms. How financial institutions exercise this extraordinary privilege profoundly shapes the type of economic and social plans that are enacted, deciding what sectors and ventures live or die, expand or stagnate.[4]

Acting as a vast planning machine, linking past, present and future, finance is directing us towards a world of accelerating breakdown by investing heavily in carbon-intensive activity. Since the 2015 Paris Agreement, thirty-three global banks have financed the fossil fuel industry to the tune of $1.9 trillion through lending and underwriting.[5] The scale of support, through equity investment and direct lending to corporations, is no doubt encouraged by the estimated $370 billion of subsidies given by governments to fossil fuel companies each year – over three times that granted to renewables.[6] And the fossil fuel giants are consistently failing to reshape their business model towards renewable energy: 'the proportion of total capital expenditure by the largest oil and gas companies, going to low-carbon businesses, represents less than 1 per cent of investment.'[7] The powerful asset management sector also invests heavily in carbon-intensive industries, reflecting the central role of carbon-linked assets in our economies. According to Tooze, 'one-third of equity and fixed income assets issued in global financial markets can be classified as belonging to the natural resource and extraction sectors, as well as carbon-intensive power utilities, chemicals, construction, and industrial goods firms.'[8] The carbon bubble is

vast and systemically dangerous to our financial systems
and the health of our planet. The world's largest three
asset managers, for example, held hundreds of billions of
dollars' worth of fossil fuel investments in 2019.[9] Taken
together, as of October 2019, companies had 'secured
financing from investors in the global capital markets –
worth $85tn (£67.2tn) for stocks and $100tn for bonds –
that will keep the world on a trajectory consistent with
catastrophic global heating.'[10] As a report from a couple
of JPMorgan economists admitted, absent transformation, 'we cannot rule out catastrophic outcomes where
human life as we know it is threatened.'[11]

Too Little, Too Late?

What 'green' finance there is – and it has grown significantly over time – still lacks the necessary scale or depth to
drive systemic change, held back by a disconnect between
time horizons, uncertainty, insufficient data, lack of consist-
ency or accountability, and costs. Many green bonds lack
transparency, and no effective taxonomy to define 'green'
assets has emerged. And too many much-heralded ESG
funds are failing to deliver substantial environmental or
social benefits,[12] with a majority only slightly altering their
portfolios from a mainstream financial index fund,[13] and
only moderately reducing their exposure to higher carbon
investments, while often intensifying exposure to holdings
in Big Tech monopolies, Big Pharma and real estate. From
the growing scale of investment in renewable energy and

electric vehicles, to central banks better integrating the impacts of climate crisis into their management of financial risk, important progress has been made in recent years, but given the catastrophic trajectory we are on, it is clearly not fast enough. If the Global Financial Crisis was caused by the confusion of risks against uncertainty, and the mismanagement of risk, environmental crisis is a repetition of the same flaws in our financial system, but on a magnitude that is planet-threatening.

In the case of climate breakdown, the choice is simple: we can continue to extract fossil fuels before eventually limiting emissions through a crash reduction. A market-led transition, poorly managed and slow-paced, would drive rising temperatures and cause economic chaos, setting off a 'climate Minsky moment' where a sudden collapse in asset prices would have devastating knock-on economic, financial and environmental effects. Even worse, we can fail to curb emissions growth entirely, guaranteeing a future of compounding climate chaos that threatens the conditions needed for flourishing life. Or, mediated by a reshaped financial system, we can act now to wean ourselves off fossil fuels: a process of managed transition, investing in a post-carbon future while safely deflating the carbon bubble.

Death Drive by Design

Finance's apparent death drive is rooted in the underlying design of the financial system: a focus on short-term profit maximisation, the controlling nexus of managerial power and elite investors, and the privilege bestowed on banks and shadow banks to create and allocate credit. The crisis of 2008 emerged from this logic of underregulated private sector balance sheets, of 'too big to fail' Euro-American financial institutions. Despite modest regulatory reforms in the aftermath of the crisis, the system's architecture remains substantially unchanged – while in areas like shadow banking, it is in some ways worse.[14] The financial barons retain their outrageous power in the form of systemically vital banks, conglomerate banks, and other financial institutions with vast, interlocking corporate balance sheets.[15] Underregulated, beyond nation-state control, extraordinarily wealthy, and beneficiaries of gigantic implicit public subsidy through their status as 'too big to fail',[16] they are the engines of finance-led growth: an economic model generating 'speculative bubbles, financialization, rentier-like behaviour, and accumulation-by-upward-redistribution'.[17] And the right of private financial institutions to create and channel credit remains, embedding stark inequalities into economic and social relations.

The elastic production of money by banking and shadow banking institutions helps give capitalism its immense productive potential, but also its tendency to damaging instability. Against the claims of mainstream macroeconomic 'equilibrium models', the heterodox post-Keynesian economist Hyman Minsky argued that financial capitalism generates crises not as exceptions caused by external shocks, but rather through the inevitable exuberance generated by the system, leading to an unsustainable accumulation of credit and debt, as economies move from conditions of cautious growth through a euphoric boom to a Ponzi-like bust. As he famously concluded: 'Stability is destabilising.'[18] This same dynamic remains at the core of capitalist financial systems.

Finance, Alone

How did we get here, with finance failing to provide adequate levels of productive investment and funding environmental chaos, while spectacularly enriching a privileged few? The answer is rooted in capital's response to the crisis of the 1970s, in whose long shadow all politics is still framed – a crisis of the Keynesian state conjoined with a crisis of the profitability of Fordist production.[19] With class struggle on the rise as organised labour's power grew and the conflict over the social surplus intensifiying, capital's response was to reshape the operation of modern finance and finance's role in the economy through a series of decisive legal, organisational, managerial, economic and political shifts. The aim: to restart the motor of profit and quell the distributional claims of labour. Aided and abetted by the neoliberal state, which sought to 'de-democratise' decisions concerning scarcity by assigning questions of allocation to actively constructed financial markets, this process has been called financialisation.

Relating to 'the increasing role of financial motives, financial markets, financial actors and financial institutions

in the operation of the domestic and international economies',[20] financialisation has resulted in three key outcomes:
the growth of the financial sector in size and complexity,
including the size of balance sheets and instruments traded,
the increase in financial activity as a source of profit in the
economy without driving an equivalent increase in productivity growth, and the increasing financialisation of everyday life. Financialisation has occurred at multiple scales,
from the global to the local, from the firm to the household,
transforming each sector of the economy. Restraints on
lending and financial activities were lifted and capital mobility abolished by governments across the world, encouraged
by the US Treasury and Wall Street. Corporations were
re-engineered, funnelling ever more cash to shareholders as
corporate debt ballooned. Household indebtedness soared
and financial motives were introduced into the public sector.
And less well known, but vitally important, finance has also
leveraged nature, turning the environment into a site for
increased accumulation, often with the support of state
power and in the face of grassroots resistance.

The operation of finance has changed over time as a
result.[21] Most notably, the financial sector has exploded in
size as a share of global GDP. There has also been an exponential growth in complexity. Wholesale money market
financing for large corporations and financial institutions
has displaced bank-based finance as the key source of
short-term financing and working capital; 'repo' markets,
a form of short-term borrowing, primarily in government
securities, have grown to become the main global money

markets, where banks borrow the bulk of the money they need to make their daily outgoings. There has been an important move away from public exchange to 'dark pools', private exchanges for trading securities, while the rise of various forms of shadow banking such as credit hedge funds or money market mutual funds, the latest evolution of financial capitalism's search for new tradable asset classes, has been a critical shift. The size and power of the asset management industry has exploded and share ownership has become internationalised. Intra-firm borrowing and corporate debt have surged. Financial integration has woven the global economy tightly together. This internationalisation has vastly expanded the size of the financial sector – and the opportunities for profit – without making the system as a whole more resilient or more concerned with the needs of the real economy. Finance-led growth has led to deepening inequality, financial instability and environmental breakdown – but it has also generated spectacular wealth for its elite beneficiaries. The financialisation of our economies was driven and is sustained by class interests, shifting wealth and power away from labour towards finance capital and asset-holders. Today, the terrain of finance is a key site of both class struggle and value extraction. But it also incubates a potential for positive transformation.

The Limits of the Present

Given the locomotive force of finance capitalism, rescuing the future from accelerating environmental breakdown will require transforming its motives and operations. The tools deployed hitherto (carbon taxes, market-shaping regulation and divestment campaigns) can play a part, but on their own are not enough to drive a rapid, just transition. The proper regulation of markets can help reshape economic activity. But that will require a transformative, market-directing green industrial strategy, which we set out in Chapter Eight, focused on reimagining work.

Taxes must also play an important role in promoting sustainable behaviours by raising the cost of carbon-based activities, through applying a penalty to 'externalities' not reflected in market prices. Progressive taxation can conjoin environmental and social justice, dispersing economic and political power, enabling higher current spending by the state to support an expanded public realm, and challenging oligarchic wealth.[22] There are limits to what taxation can achieve, however. Behaviour-shaping taxes need careful design. Individual consumption taxes, for example, set at the high level necessary to effect radical change, would be regressive and deeply unjust, exacerbating inequality and straining social commitment to a shared project of transformation. Determining the 'optimal' rate of tax in relation to externalities is hugely difficult, given that price is often a poor barometer of value, especially in view of the time horizons and scales involved in environmental breakdown. To best reshape and 'green' financial behaviour, tax systems should embed core principles: promoting rapid decarbonisation and environmental repair while ensuring equity, redistributing wealth, and raising revenue for everyday spending to support the ambitious extension of the public realm central to transformative policies such as a Green New Deal.

Just Capital: Securing a Fair Transition

The precise total value of investment required to decarbonise the global economy and repair nature is impossible to forecast given the time horizons, uncertainties, and endogenous effects involved. Nonetheless, the rapid decarbonisation and reorientation of our societies to secure environmental sustainability will require trillions of dollars in investment annually for years to come; UNCTAD, for example, estimates $1.7 trillion a year is needed to finance a global Green New Deal, requiring a decisive increase in the volume, quality and direction of private investment.[23] But that will only be delivered if the power of finance is brought to heel.

The foundation for change must be the deployment of a new set of macro-prudential policy tools to re-anchor finance in the real economy and drive sustainable investment. This should start with the careful reshaping of capital requirements and the collateral regime to incentivise green and penalise dirty lending. Regulatory measures to reshape the balance sheets of corporations – financial institutions and non-financial corporations – can also challenge corporate power, much of which rests in their ability to operate large, highly leveraged balance sheets. Central banks should green their corporate bond purchasing programmes, many of which currently have a marked carbon-intensive bias, and stimulate green bonds – a market worth an annual volume of $170 billion in 2018, from first issuance in 2008 – by improving the legal and financial framework.[24]

The power of the asset management industry, which rests on control of other people's money, must be challenged through democratic co-determination of capital markets. New rules must ensure their portfolios align with rapid transition to net zero and give the beneficiaries ultimate say in how their savings are allocated. Pensions, as a major vehicle of collective ownership with a built-in long-term investment horizon, must be rewired; we set out further details in Chapter Six, with regard to democratising control of the economy. And we need to transform the insurance and reinsurance sector, an essential but often ignored pillar of contemporary finance. Tasked with insuring society against risk, their actions – by underwriting coal, oil and gas projects – are accelerating environmental risk. New rules must align their underwriting activities with a 1.5°C pathway, including a prohibition on insuring any new fossil fuel projects and divesting from extractive sectors. A financial transaction tax can disincentivise bad behaviour, slow excessive equity churn, and challenge predatory forms of capital. The proceeds could help finance environmental reparations from the Global North to the Global South, through mechanisms such as a global climate stabilisation fund and resilience fund programme – as set out by the political economist Keston Perry – which would address loss and damage in marginalised and former colonised societies.[25]

All financial systems rely on intermediaries who observe, measure and judge activity to shape decision-making. These institutions of assessment (auditors, credit ratings agencies, actuaries) consequently play a critical role in the direction of investment and operation of the economy.[26] Yet these vital evaluative tasks are primarily conducted by profit-seeking actors, generating incentives and behaviours ill-suited to the forms of judgement required. To better manage environmental risk, new measurements and standards of judgement and knowledge are urgently needed. From a public alternative to the Big Four accountancy firms, to centring environmental and social costs in accounting and auditing practices, a just transition will require new models of financial evaluation and measurement. One important step would be developing an internationally agreed definition of what constitutes a 'green' asset, based on environmental sustainability and social justice measures. The growing size of the ESG fund market – though only $52bn of assets during the first half of 2019[27] – demonstrates appetite, but different actuarial techniques and forms of calculation are needed if ESG funds are to avoid the customary pitfall of pursuing short-term profit maximisation over their purported benefits.

The Return of Planning:
Central Banks and a Planned Transition

Left to itself, the financial sector produces too much credit for real estate, financial assets and environmentally destructive activities and too little for the real economy, in particular green infrastructures, technologies and forms of enterprise.[28] To address this, central banks in coordination with national treasuries should pursue a policy of green credit guidance. By supporting financial institutions heavily invested in the mortgage market during the financial crisis, arguably central banks have already undertaken credit guidance by implicitly subsidising certain forms of financial activity over others. By contrast, green credit guidance should actively shape the supply of credit in the economy – the volume and direction – away from financial and real estate assets and towards productive, low-carbon forms of investment.[29] Green credit guidance would help reassert public authority over the financial system. Though seemingly radical, it would represent the return of a tool commonly used in Western political economies until, as part of the wider neoliberal turn, it was replaced with a market-based approach to the creation and allocation of credit. However, from Germany's public banking network's directing credit towards low-carbon projects to the successes of China's extraordinary industrialisation – guided by a constellation of strategic credit direction, a banking sector largely state owned controlled via the

China Investment Corporation, and capital controls[30] – the evidence is clear: credit guidance works.[31]

Notably, in recent years even traditionally orthodox institutions like the Bank of International Settlements and the IMF have argued that emerging markets need to adopt the tools of the 'Beijing Consensus' to manage their economies in response to increasing volatility and financial risk: action to prevent excessive currency devaluation, aggressive regulation of systemically vital corporate balance sheets, and the use of capital controls to limit excessive capital movement, limiting the scope for financial integration.[32] Given that coronavirus triggered the worst emerging-markets crisis in history, in particular record capital outflows that are devastating in the near and longer-term in a dollar-based financial system, and given the compounding volatility which environmental crisis will induce, these same tools may be necessary, adapted to the needs of specific political economies, for many more economies in the years ahead. This should reflect a wider transition from a dollar-based global financial system and international political economy that remains deeply extractive from the Global South towards the Global North, towards one that better and more justly reflects economic multipolarity.

Taken together, this would be an agenda for repurposing central banking and ambitiously re-embedding finance in society.[33] The matter is urgent. Central banks are currently afflicted by 'a severe case of collective institutional amnesia', and we suffer as a result.[34] Their independent institutional arrangements are the historically contingent result of the restructuring of economies in the 1970s;[35] their pursuit in 'normal times' of the limited goal of price stability is a deeply inadequate framework for an age of low inflation, environmental instability and compounding risk. Their insulation from pressure from below tilts their preference in the management of inflation – whose rate reflects political conflict over the social surplus – towards privileging asset-holders over working people.[36] And their reliance in 'normal times' on a single instrument, control of interest rates, to steer the economy is too blunt a tool and has lost much of its force in a time of secular stagnation. In times of crisis, meanwhile, their immense power is used to stabilise an extractive financial system. Rather than shying away from the directional power wielded by central banks, their planning function should be made visible, accountable, and harnessed to purposeful transformation by actively steering the economies they oversee towards sustainability.

A Radical Step Change in Public Investment

A financial system brought under social control is a precondition for wider systemic change. However, the simultaneous, rapid and deep transformation of all wealthy economies will ultimately require a step change in public-sector investment. The reasons for this are clear: 'nothing on the scale and speed of required investment has ever been achieved before without direct state financial support.'[37]

Public investment can scale the infrastructures and technologies the private sector will not and more affordably, undertake risky long-term investment, crowd in private money into green activities, and purposefully reshape the direction of the economy, aligned to a green industrial strategy. Leaving the pace and direction of change to intermediation by private finance cannot guarantee efficiency or equity. While central banks should play a more conscious, directional role, a commitment to the primacy of democratic power over technocracy means fiscal policy should lead, with monetary policy supporting a step change in public investment.

To that end, as economies emerge from their Covid-19–induced hibernation, an unprecedented green stimulus must be at the centre of plans for ambitious reconstruction. The moral case for achieving rapid sustainability is irrefutable; now the economic case for a huge surge in green investment so that societies can build something new out of crisis, rather than reinflate the old, is overwhelming. A transformative increase in public investment should put our societies on the path to sustainability, accelerating the expansion of infrastructures – social, physical and digital – needed to secure a sustainable future while restructuring wealth and power in favour of workers and communities. The UK's Committee on Climate Change recently estimated a total resource cost for a transition to net zero decarbonization in the UK by 2050 of between 1 and 2 per cent of GDP per year.[38] But both the capability to go further and the responsibility to act faster should compel a much more ambitious time horizon, for richer economies, than 2050. Wealthier nations should therefore introduce a transformative green stimulus – between 4 and 5 per cent of GDP as conditions demand – as they emerge from Covid-19 to drive a just recovery that locks in a rapid transition. The stimulus should be automatically renewed each year until wealthier economies are safely sustainable, full employment is secured, and the conditions for universal security and wellbeing are met, with investment targeted at frontline communities that have borne the brunt of the damages of extractive capitalism.

The extraordinary fiscal and monetary response to the economic crisis of coronavirus has demonstrated the elastic, expansive power of finance. We can mobilise the resources we need. The more pressing question is how. Given both the scale of public investment needed to transform the economy and the low cost of public debt, a step change in public investment should be met through an increase in public borrowing. Critics sometimes argue this unfairly burdens future generations with debt. Yet that is the virtue of borrowing when it comes to financing transition. The cost *should* be shared across both present and future generations, since the benefits are spread over time: investments made today will benefit society decades into the future, thus the costs should not be borne wholly by the present.[39] Long-term public-sector borrowing is therefore an effective mechanism for pooling resources and fairly spreading the costs of decarbonisation. Future generations are unlikely to begrudge slightly higher, manageable levels of public debt, if it means they inherit a habitable world.

To deliver sustainable borrowing, two important changes are required. First, fiscal rules must be rewritten for an era of breakdown. These should allow for additional public borrowing for investment where the benefits outweigh the cost (overwhelmingly the case for green transition), and penalise governments that underuse their spending power.[40] Inflexible rules focused on public debt and borrowing limits are dangerous given the environmental crisis demands large-scale, long-term public borrowing to finance the scale of investment needed. New rules – such as the introduction of a government net worth target that acknowledges the value of investments made – can unlock the power of public balance sheets.

Second, a new compact between fiscal and monetary authorities is required to keep long-term public borrowing costs low, thereby enabling a sustainable expansion in public investment. With borrowing costs from private lenders at record lows, governments should take full advantage. Rock-bottom interest rates, coupled with the fact that debts do not need paying back in the near term, means the total level of public debt can be brought down over time through effective public debt management, with a combination of economic growth (based on lower material throughput) and moderate inflation gradually reducing the debt-to-GDP ratio.

In two moments of extraordinary financial pressure – the Global Financial Crisis and the corona conjuncture

– there has been little evidence of a sharp rise in the cost of borrowing for governments in wealthier economies, despite historically large peacetime increases in debt levels. Indeed, there has been instead a continued secular trend towards low or falling interest rates, which a global economy beset by overcapacity, weak demand, structural imbalance and an ageing population is likely to accentuate. But, if public borrowing costs do rise sharply before a green transition is complete, monetary financing of a green stimulus should be considered. Central banks should buy public debt in the form of government-issued bonds, thereby driving down the cost of borrowing for governments and increasing the scope for higher rates of public investment financed by an increase in sustainable levels of long-term public debt. Central banks would use their powers to create money to buy government debt and hold it permanently on their balance sheet, instead of being repaid by governments. Exceptional as this may seem, a form of monetary financing has already been in operation since the financial crisis of 2008–09. Both the Federal Reserve and the Bank of England have bought government debt through quantitative easing, thereby funding government spending, and though technically the money is supposed to be paid back by the government, in practice this is very unlikely, and given the example of Japan's experience in managing public debt, potentially unnecessary. As we emerge from the Covid-19 emergency, a combination of effective public debt management combined with innovative monetary financing, in exceptional and limited circumstances, should therefore finance a transformative green stimulus.

Prevailing macroeconomic conditions – low inflation, weak aggregate demand, the ultra-low cost of borrowing, and the ability of the UK and US governments to issue bonds to cover the cost of spending – makes the moment ripe for an ambitious expansion of public investment.[41] What's more, the secondary benefits of increased investment far outweigh the costs, from social and health gains to positive economic spillover effects and the creation of sustainable, equitably held wealth. In other words, a step change in public investment is not only affordable, it is beneficial and fair. Now is the time to act. Indeed, failing to take advantage of prevailing macro conditions would be an act of fatal complacency, in that delay only increases the future cost of future action; and the price of inadequate action is deadly. Equipped with the power to increase public investment, national and local governments must invest with ambition to build the infrastructure and institutions of a future of sustainable plenty.

Twenty-First-Century Public Banking

An ambitious concordat between fiscal and monetary authorities can lay the foundations for a new generation of mission-oriented, publicly owned investment banks. Their goal: to support a rapid and just transition, by lending to green businesses and financing green infrastructure projects, derisking investment in innovative new technologies and infrastructures; to pursue regional rebalancing; and to drive industrial transformation and the scaling of the democratic economy.[42] A national investment bank, supported by a network of regional divisions, can support a green recovery from the economic trauma of coronavirus.

Initial capital should come from a country's Treasury, with the banks leveraging their capital base to grow their balance sheets and increase lending by issuing their own bonds, guaranteed by their national government and central banks, as economists Laurie Macfarlane and Christine Berry have argued. As a long-term, low-risk investment, they are likely to have ample interest from pension funds. Public banking should seek to achieve returns on capital at least equivalent to the capitalising government's medium-term cost of capital, but should not be maximising profit, given its prioritisation of social and environmental goals. Admittedly, if investments are required to be bankable, capable of delivering positive long-term returns across the investment portfolio, there is a risk that status quo measurement techniques may jar with environmental and social goals, and that environmental risk proves incommensurable with financial risk. If so, new measurements focused on managing environmental risk are required; an urgent task in any case, given the new era of risk that breakdown has created.

Alongside public investment banks, we should also build an alternative to the oligopoly of large, investor-owned retail banks. The striking failure of the big banks to deliver loans to small businesses during the coronavirus crisis, despite explicit underwriting of the risk by the state, reflected their focus on more profitable activities, not the vital utility functions of a financial system: processing payments, channelling savings towards productive investment, and acting as an intermediary to transfer resources across space and time. A new generation of public retail banking can provide the vital function of finance, investing returns in improving services and accessibility while building on the growing movements for banking reform in the UK, US and elsewhere. Alongside this, a network of publicly funded and owned venture capital funds should be created that would aim to accelerate innovation at the frontier and invest in high-risk, high-reward ventures.

Finance vs the Future

Renewing hope requires transforming the future finance has planned for us – a world of growing climate apartheid, ecological trauma and avoidable social pain – by reasserting the capacity of politics to reimagine social and economic life.[43] That must start by remaking our financial institutions.

Dismantling the citadels of rentier finance and bringing investment under social control will not be easy, and will prompt a powerful counter-reaction. Responses must be prepared.[44] But it is possible. Transforming finance will require a dual strategy, a war of position *and* a war of manoeuvre,[45] reform alongside rupture, together capable of extending social control over investment in the economy. It will require a new conception of political time and action: the terrifying novelty of ecological catastrophe within our lifetime requires transformation on an unprecedented speed and scale.[46]

We cannot afford to let another crisis go to waste.[47] The Covid-19 emergency will decisively and fundamentally restructure our economy. Unlike the Global Financial Crisis, when extraordinary action was taken to shore up rather than remake a failed model, the conjoined emergencies of public health and environment must be leveraged into systemic change of our financial system. If we fail, the harms of breakdown will accelerate and multiply. Under democratic control, the immense power of finance can help build a future fit for life.

6

OWNING THE FUTURE

We want to deprive the capitalists of the power that they exercise by virtue of ownership. All experience shows that it is not enough to have influence and control. Ownership plays a decisive role. I refer to Marx and Wigforss: we cannot fundamentally change society without changing its ownership structure.

Rudolf Meidner,
interview with *LO* magazine (1975)

On 10 May 1869 at Promontory Summit, Utah, Leland Stanford brought a hammer down on a 17.6-carat golden spike, driving it into the earth. With this action, Stanford, one of the great American industrialists, had completed the ceremonial 'last spike' connecting the tracks of the First Transcontinental Railroad, which now stretched unbroken across the continental United States of America. A painting by Thomas Hill captures the moment: the embryonic power of American empire, the confident, white masculine culture of the nineteenth-century industrialist, a sense that the wilderness was a space to be mastered and dominated. Who is missing from the scene is just as telling: the migrant labourers whose backbreaking work had connected a continent, and those whose land and livelihoods had been violently dispossessed by the iron girders and carbon power of the Industrial Age and the locomotive force of fossil fuel capitalism.[1] So, by what means had this vast continent been measured, quantified and transformed?

The key to this is the nature of one of the pivotal institutions of the age of environmental breakdown: the corporation. The development of the railway networks in the United States and Europe were among the largest industrial projects of the nineteenth century. Markets, financial instruments and the modern banking system grew to speculate in railroad companies and finance their activities. By the end of the century, railway companies dominated the New York Stock Exchange.[2] What made this possible was a rapid increase in coal production with its portable, metabolic energy, and, critically, the corporation, a new institution of world-making power.[3]

The modern corporation is a legal vehicle to structure capital investment, endowed with extraordinary privileges to organise production for profit. A political, legal and economic entity, it emerged as a way of 'building technical-spatial arrangements (initially colonies, canals, and railways, later oil fields, dams, urban fabrics, industrial processes, and consumer worlds) whose scale, durability, and powers of control promised a future stream of income that could be traded speculatively in the present'.[4] This remains the core of the corporation: an institution of hierarchical coordination for managing financial flows between the past, present and future, driven by logics of competition and accumulation. As an organisational and managerial form, with separate legal personhood and limited liability, its ability to organise labour and capital over time and space has created vast wealth, connected continents and transformed and brutalised ecologies; as an authority organising production, the firm is oligarchic and unequal, with shareholders and their managerial agents sovereign. The corporation's power has then reshaped social relations and the Earth's natural systems. We live, to a large extent, in the world it has made: its hierarchies and inequalities, its poverty amid great wealth and productive power, its propulsive force hurtling us towards breakdown.

The corporation is the original and vital public–private partnership, a central institution of capitalism's infrastructural power, enabled and sustained by public action. The coronavirus crisis has, like so many before it, underscored this co-dependency and the inseparability of the economic from the political. From bailouts to wage subsidies to the backstopping of money markets by central bank action, the entanglement has never been tighter. But, instead of the public interest reclaiming enterprise as a generative and shared endeavour, the advance guard of the 'shock doctrine' is mobilising: private equity and hedge funds are poised to hoover up assets on the cheap as a whole generation of SMEs risk going under; the 'Amazonification' of whole sectors is gathering pace; and rentiers are benefiting from public funds as workers and renters bear the brunt.

The result: if nothing changes, we will see a vast upwards transfer and consolidation of wealth and power. Under such conditions, we cannot hope to secure a just transition. We need an alternative, a vision for a pluralistic economic commonwealth: a thriving ecosystem of business forms, where ownership is held in common, governance is democratic, and purpose is rooted in serving needs.

The Corporation Captured

Since the 1970s, an entwined series of legal, managerial, financial and organisational changes have transformed the corporation from an institution focused on production – even if one still laced through with injustices – into an engine of increasing wealth extraction, growing financialisation, and stark inequality. Corporations have shifted from strategies of 'retain-and-reinvest', where companies used corporate earnings to grow their productive capacity and increase real wages, to an extractive regime of 'downsize-and-distribute', where the marginal use of both corporate profits and borrowing is overwhelmingly directed to rewarding shareholders and executive management.[5]

Shareholding has consolidated and internationalised,

driven by the rise of powerful institutional investors. The doctrine of shareholder primacy has prioritised the interests of asset owners over all other corporate stakeholders. New regimes of hierarchical management have emerged and organised labour has been smashed. The financialisation of the corporation has embedded financialising logic at the heart of corporate behaviour.[6]

The simultaneous growth and concentration of assets in the hands of increasingly few investors has consolidated the immense power of the asset management industry. Mergers and acquisitions have created dominant oligopolies in key sectors, stifling useful innovation and deepening oligarchic control. Managerial power has expanded. Labour has been subject to a relentless squeeze on wages, dignity and security in order to boost short-term profits, with the ability of workers to collectively organise hemmed in and attacked, and real wages falling behind economic growth. And the growth of private equity has loaded many corporations with high levels of debt and weak balance sheets.

These changes have transformed the corporation into a machine whose primary purpose is funnelling ever more cash to major shareholders and institutional investors. In the UK in 2019 alone, dividend payments from FTSE100 companies hit a record £110.5 billion,[7] more than doubling over the 2010s, a decade where workers endured the worst ten-year period for real earnings growth for over two centuries.[8] In the US, the scale of wealth extraction is even more extreme: 'the 465 companies in the S&P 500 Index in

January 2019 that were publicly listed between 2009 and 2018 spent, over that decade, $4.3 trillion on buybacks, equal to 52 per cent of net income, and another $3.3 trillion on dividends, an additional 39 per cent of net income.'[9] The increasingly financialised use of corporate cash has left business investment rates sluggish and productivity growth anaemic. The explosion in corporate debt – often used to funnel borrowed money to investors rather than for investment – has thinned out corporate resilience. In other words, the transformation of corporate behaviour has not been about benefiting workers, improving the performance of companies themselves, or better meeting social and environmental needs. Instead, it has been a project of class power, driven by and in the interests of managerial power, elite shareholders and institutional investors.

Beyond Economic Extraction

Extractive dynamics are currently hardwired into the company's operation. This is not just capital claiming surplus value created by labour, both waged and unwaged, but wealth extracted from the natural commons. Environmental breakdown and stark inequalities are inseparable from the current design and aggregate performance of the corporation, its logic of expansion – of competitive pressure to generate ever-expanding turnover and profits – locking it into behaviours that cause deep harm.[10] It is the corporate form that organises the extraction of fossil fuels and the manufacture of carbon-intensive infrastructures and technologies, that builds out environmentally and socially destructive supply chains, and brings to life economic plans that all but lock-in environmental destabilisation. In its drive to expand turnover and profits, the corporation has then generated vast increases in material throughput and carbon emissions which have devastated natural systems.

The world's economic behemoths – including fossil fuel majors and platform monopolies, Big Pharma and major financial institutions, mining giants and arms manufacturers – will not lightly abandon a business model that has hugely enriched their largest shareholders.

A reticent liberal political economy is too formalist, too blind to the unfreedoms generated by capitalist property relations to challenge these behemoths and drive the transformation required; its policy toolkit narrow and operating at too slow a pace, its conception of public reason too limited, to provide the space for the democratic energy needed in a moment of crisis.[11] Building a more generative ecosystem of enterprise will require mobilising social power and building alliances to displace the rentiers and financial barons of the fossil fuel age. After all, the political crisis of environmental breakdown consists precisely in knowing the technical solutions needed to become sustainable rapidly, yet not being able to overcome public inertia and the resistance of entrenched vested interests opposed to change. It will require an institutional turn rooted in extending social control of the economy to rescue our futures.[12]

Beginning with the corporation, this must be anchored in a deep institutional turn in ownership, governance and control towards democracy, sustainability and purposeful collective creation. Such a process could rescue enterprise. Business can be extractive and exploitative, or it can be generative, solving needs and meeting desires. A new ecosystem of ownership and control can transfer wealth and power from the hands of the few to the many and rewire economic institutions to give people and communities real, genuine agency and control over the critical decisions that impact their lives and their environments.[13] Vitally, it can also recover entrepreneurship as a social endeavour for sustainable creation.

Learning the Lessons

Transforming enterprise has a clear precedent: neoliberalism. Unlike traditional liberalism, neoliberal statecraft, though complex and varied over time and place, rejects the belief that markets emerge organically. Instead, neoliberal governance has sought to actively create market subjects, relations, and institutions, harnessing state power to extend and encase market relations and shield them from democratic intervention.[14] From the privatisation of public assets to the active financialisation of the corporation, from dismantling collective bargaining to the retrenchment of public goods and services, neoliberal policy has sought to shrink the space for potent, democratic collectivities to form and negotiate our futures in common. In insulating the economic from the political, neoliberalism neuters democratic power, shielding unequal market relations from the uncertainty and vitality of political life. Yet it is exactly that quality – of democratic vitality and collective reimagination – that is required.

Reimagining how production and exchange are organised will require a similarly conscious, restructuring state and transformative political ambition – albeit one focused on extending democracy and sustainability. If, for example, the Green New Deal is a project to reimagine the institutions of the fossil fuel age for a future of shared green plenty, at its core must be the purposeful democratisation and decarbonisation of enterprise repurposed towards meeting social needs. This may appear a radical intervention, though necessarily so against the dangerous trajectory of the present. Yet this misunderstands the issue. We already intervene and decisively shape which forms of enterprise thrive and to whose benefit. Capitalist markets and their institutions spring from deliberate, often violent and racialised, forms of state intervention rather than from spontaneous, decentralised social coordination.[15] As Sanjukta Paul argues, the public already allocates economic coordination rights in notionally wholly private economic spheres – but does so in ways that currently work against the possibility of democratising production and provision through forms of economic democracy.[16] In doing so, the law and policy generates, sustains and safeguards hierarchical and unequal distributions of political–economic power, makes market-mediated inequalities appear as a natural outcomes, and erodes the scope and capacity of democratic power to order our common life in ways that support mutual flourishing.[17] The vital question is not whether we should intervene – we already do – but whether our interventions are generating good or bad outcomes. Given existing arrangements are generating grotesque inequalities and collapsing natural systems, the case for rethinking the firm is common sense.

Reclaiming the Company

A deep institutional reimagining may appear radical given the veneer of naturalness cast over our economic arrangements, an ideological effort that projects as fixed and permanent institutions that are plastic and malleable. Yet neither the firm nor the economy is a predetermined, 'natural' institution, but instead is constituted by politics and law, where rights and powers are publicly granted, legally defined, and capable of being transformed. The market is not a space of private contract and property that precedes social action, but rather one made possible by public power, both a product of and subject to democratic intervention and reordering. As Michelle Meagher, author of *Competition Is Killing Us*, argues there is not one inevitable 'market', but many market possibilities, depending on how the rules are defined and the resources with which participants are endowed. These can be redefined and reallocated to hardwire sustainability and shared prosperity into the economy, democratising governance rights and decarbonising economic activity. Far from a Hayekian institution of spontaneous market ordering, the rights and privileges of the corporation are made possible by state action and therefore re-codable. Instead of being ruled by capital, we can remake it as a generative institution of the commons: purposeful and democratically governed, with all its stakeholders having a stake and a say in wealth created in common.

The Covid-19 public health emergency underscored the co-dependency of the private and public sectors. With major corporations pleading for bailouts to stay afloat and the financial system stabilised by an extraordinary injection of liquidity from the world's central banks, the pandemic laid to rest (once again) the myth of a clear division between them, and dealt a further blow to the tired ideological binary of private efficiency against public wastefulness. It has also, with painful clarity, exposed how all societies are deeply entangled with non-human life. Capital's dominion was powerless before a microbe. Due to environmental breakdown, among other disruptions, companies will almost certainly require further state support on a vast scale in the near future, adding fresh urgency to the need to reimagine the corporation as a publicly sustained entity organised for the public good.

Challenging Capital's Oligarchy

Central to this must be challenging the oligarchic control of the corporation over the firm. The corporation is the legal

structure that coordinates capital and labour investment in
the firm; the firm is the economic organisation of the business, a larger, more complex entity than the corporation,
involving deeply political relationships of hierarchy, obli-
gation and power. Yet the corporation rules the firm, setting
its strategic direction and determining in whose interests
the business is run. Critically, the corporation is ruled
exclusively by its capital investors: its shareholders and
their managerial agents. The company's executive – the
board – are appointed by its legislature: its shareholder
body. Capital is sovereign in the government of the corporation, its control oligarchic in nature, based on wealth via
shareholding and managerial power; labour is excluded
from the government of the company. With economic
coordination rights within the corporation, and within
economic life more generally, assigned almost exclusively
to capital, a fundamental institution of capitalism stands in
tension with economic and democratic justice.[18]

Yet the corporation is not simply a private nexus of
contracts, an institution of voluntary and discrete associations whose actions should be shielded from policy intervention.[19] It is an incorporated body that brings together a
range of stakeholders (including capital investors, labour,
suppliers and customers) for the purpose of enterprise
within a web of relationships that are far more than just a
series of discrete contracts.[20] In this process, shareholders
are scarcely the only investors, and indeed are considerably
less central to production and the company's success than
labour. This is all the more true given the changing nature

of the contemporary capitalist company and the production of value. The decline of 'Fordist' models of hierarchical production has placed a premium on 'human capital' and the intangible assets 'owned' by employees, arguably more so than on physical capital, while the rise of 'asset manager capitalism' further erodes the idea of entrepreneurial investors uniquely exposed to risk who require a monopoly on control rights to invest. Yet labour remains unjustifiably excluded from the firm's government.

Nor are shareholder rights absolute. The corporation, critically, has separate legal personhood. Shareholders do not and cannot own the corporation. Instead they own part of the company's capital through their shareholdings, a bundle of rights, including the right to receive a proportionate share of the company's profit when dividends are declared and distributed. But their rights are not total and should not generate a monopoly on control rights.

The extraordinary privilege granted to shareholders – limited liability – shields them from 'liability for the actions the corporation took on the shareholder's behalf' and 'shifts the risks of the corporation from the shareholders to the corporation's employees, creditors, and the state'.[21] It is a form of public insurance for shareholders enabled and maintained by the state. Yet with shareholders increasingly able to socialise losses and privatise gains, from the Global Financial Crisis to the economic fallout from Covid-19, institutional investors appear more akin to rentiers than active, risk-taking allocators of capital. This is especially the case given the relentless rise of passive tracker funds as a key – and automated – vehicle for 'choosing' and holding shares. A status quo in which shareholder interests override that of all other stakeholders reflects power inequalities, not economic efficiency. If the costs of environmental breakdown are to be distributed fairly, we must urgently transform who bears risk.

Commoning the Company

We should reimagine the company as a democratic commons: a social institution with multiple constituencies who share overlapping economic and political claims on the resources of the company. At present these are unjustly enclosed and extracted, but can be reorganised to better steward the underlying resources and value of the company, distributing economic and political rights democratically, guaranteeing a voice and control rights to key stakeholders.[21] This will require rewiring the firm's constitution. Instead of assigning coordination rights based on private share ownership, ownership and governance should be democratically held and exercised, with labour and society guaranteed powerful control rights.

The company as a commons would embody the principles of democratic control that lie at the core of ecosocialism.[22] Common ownership would be matched to democratic government; those affected by a given rule within the firm would participate in its governance. Finance capital's disciplining power over the firm would be eroded; labour's collective voice institutionalised.

Powers and obligations between stakeholders would be balanced. In place of extraction, inclusion, in place of illegitimate hierarchies, pluralism in governance and voice. The capitalist firm would be replaced by the social firm: placing the social good over the demands of other factors of production. To secure that future, four steps are required: reshaping purpose, democratising governance, curbing the power of the asset management industry, and socialising ownership.

Reshaping Purpose

To transform the way companies operate, we must rewire corporate governance – the rules by which a company is directed and controlled. Instead of today's focus on maximising shareholder wealth, sustainable long-term success should be the company's primary purpose, with value equitably shared between stakeholders. Freeing companies from the tyranny of shareholder primacy can rescue the extraordinary potential of democratic collective enterprise: a collaborative endeavour to create value by sustainably meeting needs and desires.

Purpose must be built on sustainability. Company law must require directors to align company strategic and investment plans with a 1.5°C-pathway – and make them liable for company-specific environmental damage. For carbon-intensive sectors, a new special class of 'green share' should be created that would be deemed to be a majority of votes on any major issue connected to the elimination of fossil fuels in the company's production and investment plans, embedding a social veto on carbon-intensive production. Greening company purpose would fundamentally transform an institution that powered extractive capitalism.

Democratising and Decarbonising Governance

Democratising the firm is the next frontier of the struggle for a democratic society. To shift from oligarchy to democracy, Isabelle Ferreras has put forward a compelling case for economic bicameralism, 'granting the same rights to workers – firms' labour investors – as the ones held by capital investors'.[23] This is a vital step, with half of the board elected from the workforce, half elected by shareholders. But in an age of breakdown, we must go further. Alongside workers on boards, and participative work councils that reorganise work towards sustainability, we need social and environmental representatives on company boards. With a majority of board members representing environmental, social and worker interests, this would decisively place the corporation under social control and enable the prioritisation of decent work and sustainable production over maximising shareholder wealth.

Economic democracy must go further than co-determination. The shareholder monopoly on voting rights within the company should end. All workers should automatically be given full membership in their company, with voting rights. And to decisively challenge the power of institutional investors, workers as a group should be entitled to a minimum of 30 per cent of the total voting rights in their company or corporate group. Exercised democratically as a bloc, this would guarantee labour a powerful shaping role in the government of the company.[24]

Institutionalising economic democracy in the firm will require a new legal infrastructure. If politics and the economy are inseparable, law is the mediating institution that ties them together, acting as a social coding system, defining the terms of economic competition and coordination, how wealth is produced and distributed, and how inequalities are (re)produced. We cannot drive the dual decarbonisation and democratisation of the economy without a transformative legal project. Law, as Katharina Pistor has argued, codes capital.[25] It creates assets, generates value, and protects property against distributional claims in ways that weave inequality into the fabric of society, stacking bargaining power in favour of employers and asset-owners. The law helps define the terms of our encounter with the environment, not just managing a pre-existing and untouched nature, but acting, as legal scholar Jedediah Britton-Purdy argues, as a mechanism for active world-making; corporate and environmental law enables and rewards behaviours that are driving the accelerating environmental emergency.

Law then has an ideological character of its own, often protecting and promoting the status quo and its beneficiaries, in ways that can preclude its progressive repurposing, placing limits on how it can recodify and construct a democratic economy. Nonetheless, a critical legal approach – as articulated by the 'law and political economy' movement in the US – will be vital to reconstructing and democratising our economic futures. This must challenge the legal treatment of 'the economy' based on narrow definitions of 'competition' and 'efficiency' as guiding legal principles, which limits the scope for democratic participation, cooperation, and the public good to regulate economic activity.[26] In contrast to the concentration of coordination rights among property-holders, this reconstruction must disperse and democratise economic control and association. Against the naturalisation of inequality, it must open up the space for democratic ordering of and participation in economic and social life. Instead of enabling activity that is driving us deeper into crisis, the law must provide security for all, treating ecological and social functions and needs not as external to the economy, but as central to its operation and purpose.

Challenging the Power of Asset Management

We cannot transform how companies operate – and hence their impacts on the environment – without taming the power of the asset management industry. Controlling vast pools of money, the $74-trillion asset management industry increasingly dominates corporate shareholding and collectively exercises extraordinary levels of control on company executives, strategies and behaviour. Crucially, asset managers, institutional investors and banks exercise this colossal power through control of other peoples' money, much of it wealth accumulated through the deferred wages of ordinary workers.[27] Vast piles of worker capital – pensions, life insurance and mutuals – currently deployed in ways that escalate the crises could be repurposed to finance a just transition. Based on the simple principle that votes in the economy should be controlled by the ultimate saver, not by financial intermediaries, the monopoly enjoyed by shareholders and institutional investors over voting rights should end. As the company law academic Ewan McGaughey argues, asset managers should be banned from voting on other people's money unless they are following clear instructions. Sectoral worker pension plans should be taken in-house, managed for the benefit of workers and the wider environment, not the interests of the City of London and Wall Street.

Transforming how our economy operates requires redesigning how our pension system works.[28] As Christine Berry argues, the pension system is no longer fit for purpose: 'highly financialised, highly privatised and highly marketized'.[29] New strategies are needed to provide a secure income in retirement to people and ensure that income derived from pensions investment is socially and environmentally just, particularly if the future is low-growth in terms of returns, either due to deepening secular stagnation or the need for companies to operate within environmental limits. And in place of private, market-dependent relations, we should scale new ways of owning, investing and governing our common capital, requiring funds to invest in socially useful activities, from zero-carbon housing to a green energy system. While the power of organised finance places real limits on the possibility of 'pension fund socialism' for now, there is a genuine opportunity that worker capital reclaimed by a strong labour movement could bolster efforts towards wider economic reconstruction.[30]

Owning the Future

Reshaping purpose and restructuring control is not enough. Ownership is power. The gravitational power of ownership determines the contours of any political-economic system.[31] As Thomas Hanna and Andrew Cumbers argue, ownership of the means of production 'underpins all other societal values and interactions, including our relationships to each other, to work, to the rest of the world, and to nature'.[32] At the firm level, worker ownership can redistribute income and control rights, giving workers stronger democratic control over their daily lives and a share in the fruits of what they produce together. Strategically, it can begin to break the controlling power of private capital in decision-making, both at company level and across the wider economy. Socialising capital at scale can transform private and corporate wealth into public, deconcentrating capital and ensuring a more equal distribution of wealth, income and control.[33] Critically, it would enable the interests of all of society – not just shareholders and asset managers – to shape economic development towards sustainable, equitable ends. An agenda for 'funds socialism' is required: the socialisation of financial assets into funds owned and controlled by workers or society, providing a pathway to a form of market socialism.[34] To transform company owner-ship, firm-level worker ownership funds should be estab-lished for all large firms, with a growing share of political and economic rights vested in worker-controlled funds, transferring a portion of wealth and power from external investors towards the workforce as a whole. Sectoral wage-earner funds could complement firm-level funds. Modelled on the Meidner Plan, a plan devised at the high-water mark of Swedish social democracy, companies would be required to issue profit-related payments from firms in the form of voting shares to sectoral funds, controlled by workers from across the industry, socialising wealth and control over time.[35]

To democratise capital at scale, a network of national and regional social wealth funds should be created. Owned directly by all the people of a defined region or place, the funds would guarantee the public a share in the common capital stock, challenging inequalities of resource and control in the economy, transforming private wealth into equally shared public wealth, and ensuring that returns on capital are more equally shared across society. Operating independently but with a mandate to invest in the just tran-sition, social wealth funds can play an important role in ensuring companies better meet ambitious environmental and social goals.

The moment for a transformative ownership shift is now. The economic effects of coronavirus have sent share prices and public borrowing costs down to record lows. Governments should take advantage, issuing new bonds and using the cash raised to purchase a broad range of assets to be managed by the funds.[36] By buying equity and other assets at substantially reduced prices, via public debt-financing while borrowing costs are low, social wealth funds can help restructure economic activity, act as a counter-cyclical economic stabiliser, and secure a windfall for the public from their investment.

Coordinating shifts in ownership should be the People's Asset Managers (PAM). This strategic body, named by the economist Grace Blakeley, would use the tools of the finance sector to socialise ownership. As Blakeley argues, 'if institutional investors like Blackrock, who manage billions of dollars' worth of other people's assets, have become some of the most powerful entities in the international economy, then the creation of a democratically-owned and run alternative could be a revolutionary project'.[37] Acting in concert with national social wealth funds and national investment banks to identify investment opportunities that would 'promote collective ownership over strategic sectors of the economy and increase investment in socially and environmentally desirable activities', the PAM could also 'manage the private assets of domestic savers via public pension pots, and the mutual and insurance funds that currently send their capital to private asset managers for investment'.[38] These funds should be encouraged – whether by regulation or tax incentives – to pass their assets to the PAM for management, which would seek to generate risk-adjusted returns. This would provide a large and growing pool of capital that can be directed towards sustainable investment and shareholding practices, not the enrichment of the asset management industry and major shareholders.

Scaling Alternative Models of Ownership

Transforming the corporation is not enough. We require a flourishing, pluralistic landscape of place-based and purpose-driven enterprise. This is all the more critical when coronavirus is poised to decimate a whole class of small businesses, with private equity hoovering up distressed assets and consolidating the corporate sector. We stand on the threshold of a devastating wave of economic concentration unless we act to nurture an alternative.

There is an abundance of organisational forms that combine worker voice, economic equity and sustainability. Mutuals, cooperatives and social enterprises embody a different architecture of ownership and power, rooted in democratic control and the retention of common wealth in the places and communities where it is produced. Critically, they are better equipped to place serving social and environmental needs over the maximisation of profit for external investors. Employee-owned companies, community land trusts, and 'B' corporations also operate with different rhythms to a conventional corporation. Pursuing purpose over profit maximisation, they invert the traditional dichotomy of conventional capitalist companies: labour hires capital, but retains control.[39] Indeed, a critical point stressed by the sociologist Erik Olin Wright is that many forms of business in our economy are already not 'capitalist' in the critical sense of being ruled by capital, with other stakeholders subservient, organised to maximise accumulation for property-holders.[40] The challenge is to nurture and expand them.

Yet such alternative models of enterprise currently operate in a hostile economic environment. Our laws, regulations and financial system are geared to serving the profit-maximising corporation; all other organisational forms are disadvantaged, siloed, held back. If we want to see them expand, goodwill is not enough. An alternative landscape of enterprise to the status quo, one anchored in just production for the common good, will require restless institutional experimentalism. We will need new forms of accounting and auditing that better measure value and account for social and environmental needs. New types of finance, better suited to the needs and time horizons of democratic forms of enterprise will be key if they are to thrive.[41]

The Return of Public Ownership

The final element required for a pluralistic economic commonwealth – sustainable and democratic by design – is the return of an institutional form long proscribed under neoliberalism: public ownership. For, just as public ownership and investment were key to the twentieth-century developmental states of Europe and the US, helping build the infrastructures and technologies that underpinned the prosperity of the post-war era and drove the explosive growth of fossil fuel capitalism in that period, so twenty-first-century public ownership can help lay the foundations of a prosperous, just, and sustainable economy.[42]

In public hands, at municipal, regional and national levels, vital services and utilities can be organised as a right, not a commodity. Any surplus generated can be reinvested to drive rapid sustainability in the basic infrastructures we all rely upon, rather than disgorged for external shareholders. With a wave of (re)municipalisations taking back control of public services in cities and towns across the world, democratic control is returning to the stage of history.[43] A new consensus is emerging.

Given the intersecting crises bound up in behaviours of the profit-maximising corporation, our ambitions should not rest with traditional sectors. A twenty-first-century vision for democratic public ownership involves reimagining the emerging commanding heights of the next economy: digital infrastructure, from full fibre networks and cloud computing to the natural and common resources that are critical to the continued functioning of our planet. The pandemic has exposed the limits of a market-led approach; delivering universal access will require treating foundational social and technical infrastructures as public utilities, organised for people not profit. And that must be matched to a new approach. In place of the traditional, top-down, managerial forms of public ownership that were widespread, new ownership structures should be decentralised and democratic by default, and transparent and accountable where more centralised forms of coordination are called for. An emphasis should be placed on democratic governance, participation and production that is shaped by the knowledge and capacity of workers, users, and citizens to meaningfully influence workplace and enterprise decision-making.

Energy Democracy

In the US and Europe we may need to nationalise and dismantle the fossil fuel industry, both rapidly and fairly. Taking the largest oil and gas companies into public ownership would enable the overhaul of their mission and investment strategies, enabling a just, managed decline of fossil fuel production that safely deflates the dangerous carbon bubble. With their fixed capital converted or written off, the companies could be transformed towards renewable production where possible. Purchasing a controlling stake could happen either through bond issuance or the deployment of quantitative easing; given the precipitous decline in the value of many of these companies, already underway, but accelerated by the impacts of Covid-19, this is likely to be a less costly option than previously assumed.

Indeed, it is a moment of acute vulnerability for the sector, one that can be leveraged for systemic change.

The logic is simple but transformative. Regulatory fixes could attempt to change energy companies' behaviour, or drive near unprecedented shifts in capital allocation, but given the stakes and time horizons involved, it is highly risky to bet on a strategy that has hitherto failed decisively to deliver the necessary changes. While markets are moving – the share price of major fossil fuel producers is lower than it was half a decade ago and asset management behaviour is changing – this is not happening quickly enough, nor shifting into renewable alternatives at the scale required. Nor are they investing to rapidly diversify.

We cannot, then, rely on the sector or the financial system to safely manage its assets. Nor will it prioritise a just transition for workers and frontline communities. Capitalism is bad at equitably managing major energy or industrial transitions. Absent state intervention and guidance, the short-term incentive to maximise production and profits will overwhelm the need to keep the vast majority of fossil fuel reserves in the ground. While bringing the majors into public ownership might hinder the flow of profits to shareholders, a strategy for managed decline of the sector and a worker-led reorientation of their skills and know-hows towards sustainable energy generation would give the planet a chance to revive. As Johanna Bozuwa and Carla Skandier argue:

Answerable to the public and without the growth imperative, the government would be much better poised to manage their decline by directly cutting fossil fuel production from existing and under development sites in accordance with a 1.5 degrees Celsius global heating rise limit – as well as stopping new developments that are clearly outside the carbon budget.[44]

If the choice is between the profitability of the fossil fuel companies or planetary survival, public ownership can provide a potential bridge to a sustainable future.

Unfortunately, the fossil fuel reserves held by publicly listed companies are dwarfed by non-public companies held in often undemocratic hands. Nonetheless, reshaping the purpose of existing public fossil fuel companies via public ownership would powerfully impact the functioning of the global economy, firmly signalling the need to end its carbon addiction and redirecting downstream industries. If the carbon era was built on the capture of natural resources for private profit, public ownership can transition societies across the world to different energy systems – and economies.

At a more local level, community, municipal and cooperative ownership of energy must play a central role in rapidly scaling a renewable-energy network, learning from the successful energy transitions in Denmark and elsewhere. From generation to distribution, the future of renewable energy is decentralised, democratised, and decarbonised, focused on rapid transition rather than short-term profit maximisation. Public policy can play a vital role – from clear price signalling and public funding to a publicly owned energy grid and market-making green industrial strategies – in scaling such a green energy system. With the collapsing cost of wind (both onshore and offshore) and solar energy, and new technologies coming 'online', including super-cheap batteries, floating offshore wind, electrolysis, improved heat pumps and ubiquitous digitisation, there is an extraordinary opportunity to build a future of green plenty that joins together climate and economic justice.

Recovering Life

Today, as efforts intensify to build a sustainable society, we should learn from those movements and processes of systemic change that have gone before. Any transformation of our economy in the decade ahead, albeit in radically different directions to the past, will be conditional on deep shifts in property relations, enterprise and ownership. If we are content only to tinker on the margins, we guarantee the acceleration of environmental breakdown and the congealing of dynastic capitalism and oligarchic wealth. Alternatively, in the face of systemic crisis we can respond with transformative solutions. In place of capital's rule, let us choose social control, with enterprise reclaimed as a vital commons of collective endeavour rooted in democratic ownership and governance, and purposeful action. Our challenge is to use the fierce urgency of the present to drive a deep institutional turn in our economic ordering towards equality, democracy and environmental justice. It is time we owned the future.

7
COMMONING THE EARTH

We don't just fight for constitutional rights; we fight for the right to exist.

Sônia Guajajara, Brazilian indigenous leader

The Earth is a common treasury for all.

Gerrard Winstanley, English Digger, April 1649

Towards the end of 2018, Jair Bolsonaro, a far-right populist and climate change denier, was elected president of Brazil. The Amazon was firmly in his sights. The rainforest, the most diverse ecosystem in the world, home to over 3 million species, was quickly thrown open to further mining, logging and agriculture. Indigenous Amazonian communities were subjected to an ongoing wave of encroachment. In August 2019, with the reportedly tacit approval of the Brazilian government, a coalition of agribusiness, smallholder farmers, and loggers illegally set fire to vast swathes of the Amazon to clear it for further resource extraction.[1] The fires surged through the world's largest terrestrial carbon sink and choked the skies. Over 2,000 kilometres away, a smoke cloud squatted above São Paulo blocking out the sun, casting the city into darkness.

Combining the chauvinist's presumed right to exploit with the nationalist's assertion of sovereignty, Bolsonaro's leadership embodies the age of environmental breakdown.[2] The touchstones of his rule[3] – its valorisation and reinforcement of social and racial hierarchies, embrace of colonial logics of expansion through dispossession, and disdain for democratic institutions[4] – are grim portents of rising eco-ethnonationalism, a project of biological annihilation, a death cult.[5] Yet the engine of this destruction was not a unique outgrowth of his politics, nor a horrifying exception in an otherwise peaceful world system. Instead, the perpetrators were driven by a fundamental dynamic of global capitalism: the enclosure and extraction of wealth from the commons to drive further expansion. The result: the commodification of nature, the ceaseless, unsustainable extraction of resources for consumption, and the pursuit of private accumulation regardless of wider cost.

The Amazon may now be at the edge of a devastating tipping point.[6] Deforestation and rising temperatures have put the Amazon basin at severe risk of irreversible desertification. The consequences would be catastrophic, devastating ecosystems, displacing millions of people, annihilating whole species, and supercharging global heating.[7]

Capitalism would create a desert and call it profit.

A Deepening Rift

Capitalism has privatised, transformed and exhausted many of the Earth's natural systems. It has ended 'nature', leaving almost nothing undisturbed, entangling the economic with the environmental in a starkly hierarchical, extractive relationship. The environment is treated as a 'barbarous' other, an external object to be ransacked and re-engineered for profit.

The 'Great Acceleration' – the vast increase in economic activity since the 1950s – was made possible by an extraordinary increase in energy use and resource extraction.[8] The hitherto tight causal relationship between growing GDP and material and energy throughput reflects in part the taking of value from the natural commons, the transformation of energy – both fossil and foodstuffs – into wealth.[9] This process deepens and extends the rift at the heart of capitalism: the metamorphosis of energy and materials into money in ways that generate crises in social and ecological relations.[10] Capitalism is a metabolic exchange that transforms nature and labour.[11] As Marx argued, labour is the 'process by which man, through his own actions, mediates, regulates, and controls the metabolism between himself and nature. He confronts the materials of nature as a force of nature.'[12] Extraction and exploitation is at the heart of this process: from the biosphere, from a natural world inescapably entangled in the productivist force of capitalism, and from labour. Unless we act to break this dynamic, building a new and generative metabolic relationship between nature and labour, the movement from extractivism to environmental breakdown will accelerate.[13]

The Commons and the Violence of Enclosure

From unwaged work within the household to economic surplus extracted from waged labour, from the violence of modern forms of settler colonialism to the encoding of our shared technological inheritance as private intellectual property and the corralling of social data by platform monopolies, capitalism needs sources of value and forms of reproduction it can draw upon, enclose and under-compensate to expand.[14] Without this extraction of value and concentration of wealth, often from non-capitalist domains, it would quickly grind to a halt. For example, globally women aged over fifteen undertake 12.5 billion hours of care work for free every day; the monetary value of unpaid care work annually is at least $10.8 trillion.[15] Capitalism relies on the unwaged factory of the home and wider forms of under-paid work to enable the accumulation of unimaginable wealth for an elite few: 'the world's richest 1 per cent have more than twice as much wealth as 6.9 billion people.'[16] As such, it 'systematically undervalues social reproduction like it undervalues the natural world'.[17] It depends on securing an exploitative 'cheapness' in order to expand: transforming, governing and devastating ecologies to secure 'cheap' nature, money, work, care, food, energy and lives to make the world 'cheap and safe for capitalism'.[18]

Enclosing and extracting wealth from the commons is key to this. Commons are shared resources, from nature and social resources to culture and information, open to all members of a society, held in common and not privately owned, governed by logic other than that of private property.[19] Enclosure is the process by which common or non-private resources, assets or institutions are turned into private property to enable accumulation.[20] From the violent birth of capitalism to today's integrated markets on a planetary scale, enclosure and extraction have been central to capitalism's power.[21] These processes have remade social and environmental relations, typically in gendered, racialised hierarchical and ecologically unsustainable ways.[22] They are central to the dynamic of 'accumulation by dispossession' which is fundamental to the centralisation of wealth and power under neoliberalism.[23] A systemic response to breakdown and structural inequality must therefore be based on a new politics of commoning: in place of enclosure, expanding forms of collective, democratic governance, ownership, and control of our shared resources and institutions. We need to nurture a new, overlapping commons, stewarding our shared resources and wealth, to navigate beyond crisis.

Building the Twenty-First-Century Commons

Commoning opens up a space for managing our collective resources beyond binaries of market or state.[24] As Elinor

Ostrom has argued, the commons provides a way of democratically and sustainably governing a resource or institution that is neither the centralised state nor the authoritarian power of private property.[25] In doing so, it embeds
principles of 'radical democracy, material sustainability
and egalitarianism', challenging the ways in which ownership and governance under capitalism generate forms of
exploitation and domination.[26] The act of commoning
centres social needs and processes of collective, decentralised decision-making and mutuality, enabling new collectivities to form, charting how shared assets and resources
can be governed democratically.[27]

The commons is an institution of socialist transformation because it expands and deepens democratisation to the
economic and social realms.[28] Indeed, the commons should
be seen as 'autonomous spaces from which to challenge the
existing capitalist organization of life and labour'.[29] In
place of private rule, they seek democratic negotiation and
equity in production and distribution. By commoning
resources, we can unpick the false scarcity of enclosure,
while nurturing our shared resources sustainably for the
common good. In doing so, commoning can help centre
social and ecological reproduction in our economy, prioritising just relations between human and non-human
worlds.[30]

Fundamental design principles should shape the new
commons: the democratic governance of a common pool
of resources, the need for resources to be bounded so that
it is not a free-for-all, and a defined association of

commoners undertaking governance who have clear rights and responsibilities, and a commitment to stewardship.[31] Five areas – land and housing; mobility; reclaiming the city and scaling public luxury; natural systems and nutrition; and data, digital technologies and intellectual property – can form the core of a thriving global commons.

Reversing the Oldest Enclosure

Land is the deepest and oldest enclosure.[32] The privatisation and commodification of land drives financialisation, wealth inequality, and the multidimensional housing crisis. And a combination of large-scale land holdings and vast agribusinesses are rapidly depleting vital carbon sinks. Unless we transform land management, ownership and use we cannot halt the worst effects of environmental breakdown.[33] Fortunately, alternatives are flourishing. These include communal ownership of land and models of stewardship in rural and indigenous communities (such as Māori notions of kaitiakitanga in Aotearoa/New Zealand), community land trusts, public–common partnerships, new models of cooperative and social land ownership, and ambitious strategies for public land ownership that hold assets in trust for all.

The economic response to Covid-19 presents an opportunity to accelerate these trends and further common land. With lockdown leaving a huge number of people unable to pay their rent or mortgages, governments around the world are either suspending payments or replacing private payment with public ones. As the economist J. W. Mason argues, 'there's an opportunity to transform the social relations structured by those payments. In this case, that could mean not just replacing rent payments but buying out properties, so as to replace private ownership of rental housing with public or resident ownership.'[34] The Common Ground Trust, developed by political economists Beth Stratford and Duncan McCann, is another proposal, a powerful mechanism for the voluntary, large-scale transfer of land into social ownership, helping reduce housing costs, erode unearned rents arising from the control of a scarce natural resource, and pool land as a common resource.[35]

Commoning land for use-value can challenge the nexus of private land ownership, development and debt-driven finance underpinning the housing crisis.[36] With land progressively de-financialised, planning focused on delivery, and housing guaranteed as a right, we can build a world where everyone has a home of beauty, comfort and security.[37] Combined with a 'retrofitting revolution' to upgrade the insulation and decarbonise the heating systems of the world's building stock – a key source of emissions – we can deliver housing for the many.

Mobility Commoned

How we move is killing us. Transport accounts for roughly a quarter of global CO_2 emissions, and emissions from the sector are expected to grow at a faster rate than any other.[38]

The combustion engine car – clogging the arteries of the public realm, contributing to pollution, ill-health and premature deaths, and forcing inequality into our built environments – is a foundational object of the age of environmental breakdown. It is at the apex of carbon-based inequalities: over half the energy expended globally for mobility is used by the top 10 per cent.[39] The rapid rollout of electric vehicles, consigning the combustion engine to history, should be a critical goal of green industrial strategy. But electrifying the motor vehicle is not enough. EV batteries require rare earth metals whose extraction and supply chains are ecologically fraught and exploitative, occurring overwhelmingly in the Global South. And the transition is a unique window we should not waste to reimagine our built environments by reducing car use.[40] Against the privatisation, inequalities and environmental damage of the present, a bolder alternative is needed: public-oriented, decommodified and decarbonised transport systems, supportive of the particular needs of individuals and communities.

An urgent priority is the rapid scaling of a mobility commons, one that is attentive to everyone's needs. Electric car-shares run as co-ops and publicly owned e-bike and scooter networks; new cycling infrastructures and streets recovered for walking and socialising; dense networks of trams, electric buses and urban light rail, municipally owned and free at the point of use; investment in high-speed electric rail to connect regions and nations.[41] Given the congestion and pollution of privatised travel, this new combination is likely to increase transit efficiency. Transport infrastructure for rural communities should be upgraded. Aviation demand must be actively managed in ways that include a progressive frequent-flier levy in order to reduce the sector's current projected growth. Private jets, a bauble of the carbon elite, should be given a short window to electrify or be banned.[42]

From the US Highways Act under President Eisenhower to the reshaping of cities and towns in post-war Europe to suit motor travel, politics and powerful interests drove the car's ascendancy. As Timothy Mitchell has shown, corporations actively sought to promote carbon-intensive forms of living, including private petrol-based vehicles.[43] To build a post-carbon mobility commons will require a similarly ambitious political project, but with the goal of improving sustainability, health and public value.

That same imagination and restructuring must be applied to logistics. Global shipping is a major source of emissions. If it were a country, the sector would be the sixth-largest emitter on the planet.[44] Yet the sector is lightly regulated and its emissions are expected to surge. We need a new approach: internationally coordinated industrial strategies to electrify shipping and trucking, requirements to decarbonise 'last mile' delivery, and the exploration of tools such as cybernetic planning and public ownership to rationally plan the low-carbon movement of goods. If mobility, of people and materials, is a critical front in the environmental crisis, the future must be rooted in democratic, public and social alternatives.

Reclaiming the City

The twenty-first century will be defined, in part, by the collision between mounting environmental violence and the rise of urbanisation. The epicentre of this is the 'extreme city', marked by stark inequality, stratified by race, gender and class, with environmental harms disproportionately falling on the marginalised.[45] Yet they also contain the seeds of sustainable luxury: dense, with shared amenities and sustainable ways of moving and living.

A progressive project of urban renewal – the twenty-first-century heir to 'Red Vienna'[46] – can turn the necessity of making metropolises and towns more environmentally sustainable into an opportunity to build socially inclusive cities offering public luxury for all. Car-less streets and free transit, coupled with rewilding and urban greening, will reduce heat and pollution allowing us to share our neighbourhoods with animals, plants and insects and open up new spaces for leisure and play. The management of a new natural commons in the city can stimulate inclusive participation and help dispel urban loneliness.

A reorganisation of the city based on sustainability and solidarity will place logics of sharing at its heart. Standard housing units functioning as discrete financial assets have defined patterns of consumption and practices of social reproduction for over a century. New forms of ownership, such as the 'Mehr als Wohnen' (More Than Housing) cooperative housing neighbourhood in Zurich, allow individuals to pool resources and spaces. From shared laundries to shared roof gardens, the path to sustainability is one of sociality and encounter.

New infrastructures for leisure for all ages can nurture a conception of urban life beyond work and consumption, one geared for care and mutual enjoyment, beyond a city that acknowledges only workers and consumers as legitimate occupiers of indoor and outdoor public space. High streets built around the needs of communities, with flexible tenancy agreements and a versatile workspace offering, can build a rich social life as well as diverse and inclusive local economies.

Reimagined planning and consultation processes, which currently inscribe inequality into our built landscapes, can place communities at the centre of urban design, allowing them to create new ways of navigating and enjoying reclaimed and de-commodified public spaces. With Covid-19 laying bare both the inequalities and resilience of urban life, we must take stock of these lessons and reimagine how we move and live.

Restoring the Natural Commons

Extractive capitalism dreams up ever-more elaborate schemes for geoengineering, its elite beneficiaries seeking to decarbonise economic activity while retaining their place in the unequal structures of the present. Yet natural carbon sinks already provide the capacity to help us rapidly, cheaply and safely decarbonise, without endangering health in the process. A twenty-first-century Civilian Conservation Corps is needed in every country: an army of rewilders restoring vital boreal forests, rainforests, peatlands and grasslands, nurturing habitats and biodiversity back to health, and scaling natural carbon sequestration.[47] Land for rewilding would come from spaces reclaimed from livestock grazing; the restoration of rich ecological life to manicured 'natural' spaces, now almost wholly depleted of biodiversity; and urban spaces. The benefits are manifold. Rewilding is a powerful, cost-effective way to resist both climate breakdown and the collapse of biodiversity, while providing a new range of jobs in tending and repairing environments.

Nutrition for All

Agriculture, food production and deforestation account for nearly a quarter of human greenhouse gas emissions. Yet agricultural capitalism is still failing to provide nourishment to all and decent conditions for agricultural workers.[48] As the 2019 IPCC report into land use, agriculture and food security made clear, the current food system, underpinned by private land ownership and large-scale agribusiness, poorly serves people and the planet. According to the Food Aid Foundation, 13 per cent of humanity is undernourished.[49] The challenge will grow: the UN's FAO estimates that feeding a world population of 9.1 billion people in 2050 will require raising overall food production by around 70 per cent from 2005 levels.[50]

At the same time, unequal distribution of food and artificial scarcity condemn millions to hunger while, each year, global food production exceeds the requirements of the world population.

Food justice and ecological justice can go together, but will require transforming how we produce and value food. As Raj Patel and Jason Moore have shown, the production of 'cheap food' is hugely expensive. Wage suppression, forced labour, dispossession and despoliation, and externalised environmental and social costs have all been core features of modern food systems.[51] It is no surprise, then, that the infection hotspots of Covid-19 have often been in meatpacking plants, places of insecurity and pain for human and non-human life alike. 'Cheapness', in Moore's phrase, 'is violence.'[52] As a corollary, our food systems are strikingly wasteful: 'one third of the food produced in the world for human consumption every year – approximately 1.3 billion tonnes – gets lost or wasted', enough to feed all hungry people.[53] As Mike Davis has argued, capitalist agriculture is ill-equipped to provide both nutrition and security, 'because the world market misallocates crop production (beef over grain) and fails to deliver basic income to small producers and farmworkers'.[54]

What is required are new goals for agriculture and the food economy, working towards a system where all have access to a healthy diet, agricultural workers enjoy security, and natural systems are nurtured back to health. To restore depleted soils, sequester carbon from the atmosphere, improve long-term yields and make crops more resilient to climate change, it is necessary to improve industrial farming methods and, critically, to use more sustainable, regenerative techniques – agroforestry, polyculture, no-till farming, and organic approaches.[55] Advanced techniques like bioengineered crops and drip irrigation can be managed into the process. The knowledge and care of communities at the front line of enclosure and dispossession of land must be central to this process. Many of the staple foods of Western diets were made palatable by the skill and patience of indigenous communities. Those same techniques and qualities of nurturing, of lifemaking over profit-seeking, must be at the forefront of local food security. Agricultural trade deals, too, should be rewritten to focus on reducing hunger and poverty, not eliminating restrictions on agribusiness. As UNCTAD has shown, existing agricultural deals largely benefit the Global North at the expense of farmworkers and communities in the Global South.[56] To support more sustainable agricultural practices *and* deliver food justice – from production to plate, from packing to distribution – that must change.

Countries should also experiment with 'national food services' that can help address food insecurity, waste and social isolation. They would embody and sustain forms of mutual aid and social solidarity that emerged during the Covid-19 crisis, where anyone can find affordable, healthy, sustainably sourced meals. Food has the potential of creating forums for communing across age and class in every neighbourhood – spaces that build a new sense of the public, based on mutual care and the idea that health is something we hold in common.

New approaches to agriculture and land use must be matched by new eating habits, with more plant-based diets and reductions in meat and dairy. Livestock uses a third of global cropland and is a key driver of deforestation; and it accounts for almost 15 per cent of global greenhouse gases: every '4lbs of beef you eat contributes to as much global warming as flying from New York to London – and the average American eats that much each month.'[57] Changing this calorific and carbon injustice must be part of a global Green New Deal that delivers food security for all through fair, sustainable agricultural practices. Scaling investment in synthetic meat replacements is potentially useful, but we cannot rely on 'meatshot' technologies alone; the urgency of the problem requires action to reshape consumption patterns. From incentivising low-carbon foods to restricting certain forms of advertising to requiring major public and private bodies to procure sustainably, the age of breakdown will necessitate a new politics of food and consumption.

Building a Technological Commons: Democratising Data and Digital Technologies

The commons of nature and labour are not the only materials whose value is enclosed and extracted by capital. Our common technological inheritance, a dense and layered ecology of collective knowledge and technical systems, is increasingly privatised, its wealth corralled, technological development steered by digital oligarchs. Technologies and technical infrastructures built out of public investment and the knowledge commons, that promised the decentralisation and democratisation of decision-making, are now increasingly siloed behind private intellectual property regimes and opaque corporate structures.

Technologies are hybrid, social relations as much as technical artefacts, rediscovered and redeveloped, applied at different scales, from the personal to the systemic. Digital technologies – the networked services, devices and spaces of digital capitalism – are transforming fundamentals of economics and politics: the employment relationship and work, the production and distribution of value, the nature of a commodity, and conceptions of scarcity and abundance.[58] Breathless futurism is unhelpful, however; the 'shock of the old' remains, with uneven linearities, disorienting conjunctures, and the centrality of 'mundane' technologies to everyday life.[59]

The development and use of technology and technical systems is driven by two competing nodal points: the nexus of US-led 'Big Tech', finance capital and the imperatives of US geopolitical power, and China's state capitalism, whose technological systems are now poised to provide the digital infrastructures for much of the world. Neither are the basis for a more equitable and democratic future. Through these digital technologies, our societies – our built environments, relationships, even our sleep – have been turned into a vast dragnet for the collection, analysis and monetisation of data. The whole of society has become a single digital office and factory. Technologies colonise everyday life, their development and direction by private power reproducing and deepening existing inequalities.[60] Technical systems intimately, if unevenly, shape our lives, through forms of control and connection, and restructure production and exchange. The networks of unaccountable power that flow from digital technologies imperil a basic premise of democracy: 'the idea, often invoked but rarely attempted, that the whole of the people should determine how society is run'.[61] Instead, we are nudged, surveilled, directed, all by a knot of overlapping technical systems and corporations, at once diffuse and yet powerfully concentrated.[62] The algorithmic ordering of the universal platforms, invisible but ubiquitous, coordinates and strati- fies ever more domains of life: how we move, live, work, consume and love. These world-shaping technical systems distil to a single point: vast, growing wealth and unac- countable power for the digital oligarchs of Big Tech. Amazon CEO Jeff Bezos, for example, saw his fortune grow by $24 billion during the first months of the corona- virus pandemic, to a staggering $138 billion.[63]

Another Digital World Is Possible

The ecological is inseparable from the technological. Technologies direct and reshape matter and materials for particular ends, the contest over which occupies the core of politics and environmental breakdown alike.[64] Technologies are also dialectical, reflecting the culture and dominant economic paradigms in which they are embedded. They are malleable infrastructures imbued with human purpose and conflict, capable of reimagination. In this reality lies the potential to design and build a very different future to our current trajectory.[65] If, though, we are to rearticulate our relationship with nature, and between ourselves, in a more generative, equitable fashion, we must democratise technological development and use. This should start from an understanding of how best to meet the needs and expand the capabilities of ordinary people and their environments, rather than powerful corporate or state actors, and then intentionally design technical infrastructures to serve those ends.[66]

Decisions on how we develop, design and deploy technologies are political – profoundly shaping the distribution of power and organisation of work and materials in society, and hence key accelerators of the environmental crisis. 'Technology is neither good, nor bad; nor is it neutral', the historian of technology Melvin Kranzberg argued.[67] What is developed and for what purposes is contestable, patterned by who owns the machines and intellectual property, and who directs their development and use.[68] Whether technological systems address or amplify breakdown, cement digital oligarchy or generate shared prosperity, is therefore primarily not a matter of technical capability but of politics. We have the technologies and infrastructures capable of achieving sustainability, but, unless they are organised along a different logic an environmentally sustainable civilisation will retain the injustices of the present.

An emancipatory technological project requires a new politics of technology: more sceptical, more attentive, more aware of the trade-offs involved in technological development, not awaiting technological liberation but organising for it, demanding that technical systems serve us, not us them and their owners, creating institutions that can deepen social control over technologies and redirect them towards the creation of sustainable plenty.[69] Technology cannot emancipate; as Aaron Benanav argues, that can only come through social relations reorganised for abundance. Left to Big Tech, technological development will drive widening inequalities; to thrive we need to liberate technology from capitalism.[70]

Democratic Knowledge Worlds

To build a future of shared plenty, we need to move from conditions of private enclosure to a digital commons.[71] Vital digital, data and knowledge infrastructures should be organised as a common resource based on democratic ownership. As a first step, antitrust regulation should challenge and reduce the monopoly power of the universal platforms. More ambitiously, digital connectivity should be treated as a societal right and digital infrastructure – the rollout and maintenance of fibre optic connection and 5G in particular – organised as a vital twenty-first-century public infrastructure, the foundation of green energy grids and low-carbon transport systems. As the theorist of digital technologies Jonathan Gray argues, a democratic twenty-first-century digital infrastructure would open up a more innovative and experimental future: 'Digital infrastructures currently oriented towards extraction, exploitation and profit might instead facilitate solutions to pressing societal issues, such as inequality, housing, land reform, climate change and public health.'[72] The potential is extraordinary, from the creation of national data funds[73] and collective data banks that turn certain forms of data into common property that guard against corporate capture,[74] to intervening around algorithmic systems; from reshaping platform work to socialising 'feedback infrastructures' to better plan a just transition, and exploring how data infrastructure can be remade as sites of participation, from the local to the international. Reshaping the organisation of technical systems can drive wider changes in social, economic and ecological relations, changing the purpose of connectivity, challenging exploitative platforms and algorithms, and supporting alternative ways to use digital technologies that nourish the 'progressive composition of a common world'.[75]

By commoning digital infrastructures and data pools, feedback loops from socially owned 'Big Data' can lay the foundations for digital socialism, and enable the scaling of forms of non-market social organisation and coordination in the context of automation and environmental breakdown.[76] Exponential improvements in data analytics can enlarge our capacity to plan equitably, complexly and efficiently, over and beyond profit maximisation as the organising logic of society. As Evgeny Morozov argues, socialising 'Big Data' combined with democratic control of artificial intelligence can be a source for radical empowerment in our societies.[77]

Critically, there have been arguably no truly transformative algorithmic breakthroughs spurring the current wave of AI (or machine learning) predictions. Instead, the great leaps in capabilities we are now seeing are the consequence of scale and monopoly power: Google has the ability to run the huge data centres that the algorithms for machine learning require. Such scale is inherently energy-intensive, with the use of this kind of AI being applied to often marginal social benefits. But a new ecosystem of collective rights to data, social ownership of capital, and public analytical power to analyse data at scale could repurpose existing technical assemblages towards more sustainable, enriching ends.

Beyond the Panopticon

New collective rights to data must pay proper regard to privacy, attentive to the harmful purposes data can be put to if in concentrated hands, whether public or private, particularly against already marginalised communities. Indeed, critical engagement in how data-worlds are constructed is urgent: what is and is not observed, what data is collected, who has use of it, on what terms of access. If the externalities of fossil fuel extraction are devastating natural systems, the externalities generated by data extractivism in terms of psychological, political and environmental harm are immense.[78] Just as we need to ensure carbon remains in the ground, so a digital commons must limit the capacity of both the universal platforms and the state to surveil all of life. But reimagining how data is generated and used can help democratise technologies, moving beyond the neoliberal, carbon-intensive growth model of the 'smart city'.[79]

Challenging the monopoly power of the tech giants is a critical step. At the same time, public policy should be more innovative, incubating cooperative digital platforms – with democratic governance between key stakeholders – that can provide goods and services efficiently and effectively, while enhancing the rights of both users and workers on the platform. From social media platforms to online marketplaces, to transportation and housing applications, a new arrangement of control, voice and purpose for digital platforms can transform how they operate and for whom.[80]

Scaling a democratic communicative apparatus can reshape existing media landscapes dominated by plutocratic ownership in print media, a broadcasting ecosystem of public and commercial giants, and a digital landscape dominated by the tech giants.[81] What is needed are 'new forms of communication, distributions of knowledge, and an enhanced capacity for assembly and equality in public speech'.[82] An example might be Dan Hind's proposal for a British Digital Cooperative, a common property owned collectively by all residents of the country, 'tasked with developing a surveillance-free platform architecture to enable citizens to interact with one another, provide support for publicly funded journalism, and develop resources for social and political communication'.[83] Further steps, such as a democratised BBC in the UK, can enhance a communicative and knowledge commons.[84]

Commoning Intellectual Property

Many of the most valuable assets in contemporary capitalism are not physical, but intangible products of collective human intellect.[85] A legal and policy architecture – intellectual property (IP) – grants owners of IP extensive rights, including the right to exclude use of the IP. There is some logic to this, and it is appropriate in certain contexts. However, given the vital role of public-sector funding and R&D – and the role of the wider knowledge commons – in generating new assets, an alternative approach to intellectual property rights is needed, one that challenges rentier power, better unlocks the benefits of technological advancement and public investment, and gives the public its fair share.

From public patent pools with copy-left or commons-based licensing at the regional, national and international level, to public policy changes around compulsory licensing for public purpose, a new legal infrastructure to common key forms of intellectual property can help drive the rapid uptake of green technologies and practices.[86] One step forward would be for the public to retain ownership of IP emerging from publicly funded R&D, with the IP pooled in a knowledge commons and access to the asset governed and managed for the public good. A pressing need is to ensure any Covid-19 vaccine is open licence, so it will be used to address the public health emergency not maximise the profits of Big Pharma.

A critical generator of new inventions and advances and a vital element of a future commons is the organisation and role of basic research. Given the scale and range of techno-logical advances required, and above all their mass deploy-ment, a reliance on the existing innovation economy, rooted in improvements to existing systems and tech-niques, might not be enough. Instead, there is compelling historical evidence – especially for computing, but in other realms too – that achieving true qualitative technological leaps of the kind we need requires a basic research regime with certain characteristics. These are transformative but simple: fund people, not projects; have broad but consen-sual visions and goals; and fund at a high level without interference or worry about short-term earnings or market pressures. These same social and purposeful characteris-tics – eroded by the financialisation and short-termism of modern corporations – must be embedded in a new commons-based, collectively owned regime of basic research. These approaches should be applied not just to developing cutting-edge technologies, but decarbonising long-standing production techniques: the steel and cement industry, for example, together account for roughly 13 per cent of global emissions. Technologies underpinning the foundational and mundane, as much as the extraordinary and experimental, must be reimagined.

From Enclosure and Extraction to a Democratic Commons

The Covid-19 crisis has underscored the need to rethink how the production and use of knowledge is governed and in whose interests. In Mike Davis's words:

> Big Pharma, the monopoly of monopolies, epitomizes the contradiction between capitalism and world health. Extortionate prices and proprietary patents for medicines often first developed by university and other public researchers are only part of the problem. Big Pharma has also abdicated the development of the life-or-death antibiotics and antivirals that we so urgently need. It is more profitable for them to produce palliatives for male impotence than to bring online a new generation of antibiotics to fight the wave of resistant bacterial strains that is killing hundreds of thousands of patients in hospitals across the world. Big Pharma claims protection from antitrust laws because it is the major engine of drug research, when, in fact, it spends more on advertising than R&D . . . Big Pharma, in essence, is rentier capitalism, a fetter on the emerging revolution in biological design and vaccine production.[87]

If rentier capitalism now threatens both planetary stability and public health, a new arrangement of social ownership, democratised economic power, and purposeful commoning can provide the alternative. Against logics of enclosure and privatisation that are driving our entwined crises, we must scale a twenty-first-century commons, democratically stewarding the assets and resources we all need to thrive. In contrast to the blackened desolation of a burnt Amazon and the sight of Big Pharma profiting from crisis, an expanding democratic commons can nurture spaces of ecological and democratic empowerment from which the flourishing of society and nature can spread.

8
THRIVING NOT SURVIVING

The alternatives are stark. Either we will have a future in which human beings are reduced to a sort of bee-like behaviour, reacting to the systems and equipment specified for them; or we will have a future in which masses of people, conscious of their skills and abilities . . . decide that they are going to be the architects of a new form of technological development which will enhance human creativity and mean more freedom of choice and expression rather than less.

Mike Cooley, union organiser and
co-author of the Lucas Plan, 1980

In late 1974, threatened with mass redundancies, the Lucas Aerospace Combine Shop Stewards Committee sent out an urgent call for help. Instead of forced layoffs, could production at the manufacturing firm be repurposed to meet social needs?[1] First they sent out invitations to 180 engineers across the UK, asking for advice and ideas for industrial restructuring. High hopes were dashed when only three replies came back. But then, rather than accept defeat, the shop stewards turned to their members with a transformative request: redesign the future. The result, after a year-long, bottom-up process, was a visionary document: the Lucas Plan.

Published in January 1976, it set out a plan to decisively shift production at Lucas Aerospace away from publicly funded military equipment contracts towards socially useful technologies. Many of the proposals, including heat pumps, wind turbines and solar cell technology, were ahead of their time. Nor was the document, nominated for the Nobel Peace Prize in 1979, just an exercise in abstract engineering. The Plan articulated a vision of work fundamentally reorganised: more collaborative, with design shaped by the tacit knowledge of ordinary workers, and decision-making democratically organised.[2] It recognised that technology is political, a social relation as much as a technical one, which can be democratised and repurposed just as work can be reimagined.[3] If technologies reshape the fabric of everyday life and environments, of flows of mass and energy, an emancipatory politics must reclaim their development and use to better support human flourishing. Through its emphasis on participatory design and production to meet needs, the Lucas Plan showed the way.

The Plan, though, was defeated. Management were strongly opposed, the trade union upper echelons hesitant, and the Labour Party – beyond a narrow set around Tony

Benn, then secretary of state for energy – reluctant in its support. Job losses followed. And soon after its publication, the New Right smashed the crisis-ridden impasse of post-war social democracy to cement a neoliberal politics of work anchored in inequality. We live with the consequences of the road not taken: production geared to the needs of the market, use-value subordinate to exchange-value, work organised through undemocratic hierarchies.

Yet the unprecedented economic response to coronavirus, with whole economies deliberately demobilised, markets suspended, and price systems temporarily abolished in many sectors, suddenly prised open the very questions the Lucas Plan had explored.[4] What forms of work are truly 'essential'? How can we foster working relations built on solidarity and mutual esteem? If market values are subordinate to use values, can production be reorganised to better meet fundamental needs? Can we expand leisure time without a loss of pay? What is the purpose of the 'economy', and, as an object of intervention, how can we re-engineer it? Resuscitating a flourishing future will require excavating the promise of the Lucas Plan: centring and valuing work that nurtures and sustains life, democratising technology to extend our capacity for creativity and joy, and organising production to meet the needs of all. This vision is not an impossible utopia, but it demands more than accepting the merely tolerable as our fate. The alternative is a deepening crisis of both work and the environment. Only by reimagining how we work can we restore life amid the ruins.[5]

A Crisis of Work

Work – physical and mental effort purposefully directed to achieve a result – is a vital means by which we intervene in and reshape our environments. How it is organised and to what ends reflects, reproduces and transforms social and material relations. It is world-making in scope, reshaping physical landscapes and social infrastructures, from the granular and intimate to the planetary and systemic. Work acts as a central mechanism in capitalism for distributing income through participation in the labour market; it can provide identity and reward, and be a source of individual expression and joint, enriching endeavour. But work, both waged and unwaged, is also a realm of domination and exploitation. The wage relation is the central political conflict of capitalism, a primary venue for class struggle. Deeply political, it is structured by unequal distributions of power and authority, by degrees of compulsion and uneven discretion.[6] And the wealth labour generates through work is corralled, extracted and concentrated by capital.

Despite the veneer of a successful recovery from the financial crisis of 2008 – record stock-market highs and low unemployment – the reality, for many people, is that work is not working. Economic inequality in much of the world is at levels not seen since the Gilded Age. Real income growth has become severed from productivity growth; Covid-19 risks another lost decade of stagnant – or worse – living standards, intensified by chronic under-demand for labour globally. While many enjoy 'good' work, and a privileged few receive outsized rewards, pervasive techniques of surveillance and control erode dignity and autonomy. The weakening of the traditional employment relationship and rise of insecure work undermines security, falling hardest on the working class. Corporate power mushrooms, industries consolidate and financialisation tightens its grip. Managerial power has swelled and the legal assault on organised labour continues. Gender, class, race and migration shape unequal labour markets.[7] Forced labour, exploitation and precarity are endemic in global supply chains.[8] And the crisis of waged labour is matched by a multidimensional crisis of unwaged labour, particularly care work. Social reproduction – the work of creating and sustaining societies – is starkly gendered, undervalued and exploited.[9] If capitalism extracts surplus value from waged work, it relies upon taking value from unwaged work – and non-capitalist forms of production – to survive and expand.

The Covid-19 conjuncture has amplified the crisis of work. If the Global Financial Crisis exposed the fragile nodal points of the financial system, coronavirus has starkly clarified the truly essential forms of work: those providing care and sustenance, and the provision of goods and services on which we all rely. Yet it is this foundational work that is insecure and marginalised. Instead of protecting and investing in 'essential' workers and sectors, our economies extract value from them, channelling it into 'non-essential' sectors. A fragile, uncertain recovery risks heightening already stark inequalities in working conditions. Coronavirus is just the beginning; destabilisation will only grow as environmental breakdown accelerates, violent and uneven, deepening a sense of crisis in work – but also providing an opening for ambitious reimagining.

A Systems Crisis

The crisis of work is inseparable from the sharp concentrations of power generated by constitutive features of capitalism. Relations of production under capitalism involve 'certain forms of control over the productive forces – the labour power that workers deploy in production and the means of production such as natural resources, tools, and spaces they employ to yield goods and services – and certain social patterns of economic interaction that typically correlate with that control'.[10] In particular, the means of production are predominantly privately owned and controlled; the majority of people need to sell their labour power to secure the means to live, creating class divisions that involve relations of authority, conflict and subordination between workers and capitalists. Socially constituted markets, which people enter into with sharply different resources and initial endowments, are the main mechanism for the coordination of production and exchange and for determining the use of societies' productive surplus; financial institutions allocate money-capital via newly created bank money to finance production; and capital accumulation is the primary purpose of production, oriented towards profit over the satisfaction of needs.

Through this assemblage, capitalism has generated vast, if starkly uneven wealth. It has supported the growth of meaningful, if partial, freedoms. Capitalism's productive power, moreover, can be more equitably distributed through stronger employment protections, a powerful labour movement, and universal, strong social security. A radical social democratic programme – providing the public goods which markets under-provide, regulating negative externalities, 'derisking' life through comprehensive social safety net – can secure a fairer, more resilient society. But though its excesses can be mitigated, capitalism can never be fully mastered, nor its endemic injustices overcome.

Employment relationships are defined by sharp asymmetries of power between employers and workers. Mediated through labour markets and against background inequalities in power and resources, 'the silent compulsion of economic relations sets the seal on the domination of the capitalist over the worker . . . [The worker's] dependence on capital . . . springs from the conditions of production themselves, and is guaranteed in perpetuity by them.'[11] This disadvantage enables capital to define the terms of work and extract a disproportionate share of the economic surplus. Capital enjoys a near-monopoly on economic coordination within the economy, both strategically in allocating investment and within the firm, where workers must submit to hierarchical direction. Private ownership of the means of production therefore privatises decision-making power. Freedom for capital necessarily involves non-freedom and forms of limitation for those without property. In the workplace, the private government of our lives, our capacity to self-direct our movements, actions and speech is circumscribed and controlled.[12] And the institutional arrangements of capitalism generate stark material inequalities that inhibit the realisation of political liberty, substantive equality of opportunity and outcome, and an expansive democracy. The foundational institutions of capitalism are then in tension with democratic justice and the development of an equitable commonwealth.[13]

Working for Life

We are, therefore, on the horns of a dilemma. Despite the growing capability of machines, we are nowhere near the advent of a fully automated economy. Indeed, with deindustrialisation, under-investment in labour, and the growth of low-wage low-productivity jobs, a future without intervention is likely to be less one of full automation and more of stagnation and insecurity. Securing a post-carbon future, repairing ravaged natural systems, and nurturing damaged life will require an extraordinary collective effort. Yet unless the insecurities of work under extractive capitalism are dismantled, at best sustainability might be secured at the expense of justice, entrenching the inequities of the present; indeed it could dramatically worsen existing inequalities. A just transition must reimagine work.

Work reimagined should bring to life an 'alternative vision of wealth and experiment with ways in which human labour can be employed for the production of solidarities, mutual pleasures, and beauty'.[14] Rather than asymmetrical power and coercion, it can be shaped by 'accountable procedures, open to participation and responsive to needs'.[15] The extension of social control over economic institutions can ensure people have 'broadly equal access to the necessary means to participate meaningfully in decisions'[16] affecting their lives. Reorganised as a realm of economic freedom, workplaces can be a space to pursue 'people's development and exercise of their creative and productive capacities in cooperation with others'.[17] This means centring and supporting new forms of labour and value, rooted in solidarity, care and creation. As Alyssa Battistoni writes, work would be 'oriented toward sustaining and improving human life as well as the lives of other species who share our world'.[18]

Reorganising work on principles of equality, deep individual freedom and collective empowerment, solidarity, sustainability, and democracy will require transforming the institutional architectures that currently structure work. Abolishing work as we know it will depend on a deep reordering of power, property and purpose. Unlike today, we would all have a stake and meaningful say in decision-making that shapes our work, communities and society, and share in the commonwealth. And if inequality of property and wealth distribution creates inequalities in bargaining power – the ability to 'hold out' longer, as

Adam Smith said – reimagining work will require reshaping the forces that pattern bargaining power. Three interlocking interventions are fundamental to systemic change – and the corona conjuncture should be leveraged to effect them all within rapid order.

Three Transformations

First, we need to make the meeting of social and environmental needs the core purpose of a just and sustainable economy. Reproducing and caring for thriving life, not production for endless accumulation must be the aim, centring, valuing and esteeming care and the work of the foundational economy – the provision of the goods and services we all rely on. This will grow and better value certain forms of work, although it will shrink others.

Second, the exercise of control and planning within the economy, from the strategic and sectoral to the firm level and everyday working conditions, must be transformed. In place of narrow, private power and capital's near monopoly on shaping the terms and direction of work, we need to extend social control, democratise economic planning, and reshape and pluralise economic coordination rights. The outcome: a deep structural reshaping of the terms of work and the distribution of wealth and income towards working people and communities. Additionally, the power of the platform monopolies to calculate and direct must be challenged through a

digital socialism that reclaims technological infrastructures for shared abundance.

Third, to break the asymmetry of power that shapes
how work is organised we need to weaken the link between
income and the labour market. That requires a new set of
strategies for providing everyone with 'broadly equal
access to the necessary material and social means to live
flourishing lives'.[19] Universal basic services and an ambitious social security system that guarantees a minimum
income for all can universalise security and reset the terms
on which work is organised. A new deal at work should
guarantee vital protections and rights for all workers. This
should be combined with strategies to secure a fairly
managed reduction in working time to expand leisure time.

In combination, these approaches can overcome the
power imbalance that structures work and leaves too many
insecure, address the background resource inequalities
that structure the asymmetrical terms on which work is
organised, and reshape the company from a space of
private control to a social institution in which workers and
society have a genuine stake and a say. In doing so, it can
revitalise the old labour slogan: 'Don't beg to work:
demand to live!'

Life-Making and the Foundational Economy

Care work is environmental work: low-carbon, non-
extractive.[20] If our current economy is geared towards
production to maximise profit, a sustainable future must

prioritise the needs of people, recentred on caring and nurturing life. Care work must be properly recognised, valued and rewarded. As those calling for a Green New Deal have argued, this requires us to 'reimagine social labour and human purpose, where care across scales – individual, social, and planetary – can finally be made whole'.[21] Securing justice in ecological and social reproduction go hand in hand.[22]

That will necessitate overcoming the crisis of care. Care work is how we reproduce life: 'caring for spaces and environments, caring for oneself, and caring for each other', from childcare to eldercare and social care, from sex work to cooking, cleaning to teaching.[23] As the US National Domestic Workers Alliance describes it, care work is 'the work that makes all other work possible'. Without care work, waged or unwaged, our economies would grind to a halt. Yet care workers are undervalued, exploited and marginalised. Forming the heart of the contemporary multiracial working class, their conditions are insecure and poorly paid. As Helen Hester and Nick Srnicek have argued, the mounting care crisis is driven by multiple dynamics: 'with demand for services growing at the same time that unpaid workers are entering the labour market, paid workers are facing treacherously low wages and abysmal working conditions, and the government is stepping back from public provision.'[24]

A caring, green economy must dismantle the gendered hierarchies and uneven distribution of care work in and outside the home, and lower the overall workload of reproduction. Rethinking how we live and care can create new forms of solidarity and mutual aid while reducing the resources we consume. The agenda should be comprehensive and transformative. Free universal childcare is vital, an ambitious goal of decommodification that can attract a broad coalition behind a structurally transformative demand. A universal, generous child benefit scheme should be introduced everywhere it is absent and social security systems should work to unpick entrenched, unequal norms of family and care. Adult social care must be freed from the grip of private equity and other predatory forms of capital, provided instead through public and cooperative provision. Socially necessary unwaged work should be 'visible, valued, and equitably distributed'.[25] New approaches to redistribute work within the household, such as eliminating workplace penalties for part-time and flexible work, should be pursued.

Past crises have irreversibly transformed social reproduction.[26] Our response to environmental crisis should be to institutionalise an emancipatory politics of care. Policy must lean against the naturalisation of reproductive labour and challenge the gendering of certain forms of labour.[27] We should experiment with cooperative childcare models and community-based systems of care where everyone would be expected to contribute. Camille Barbagallo, for example, has argued for a new generation of Community Care Centres in place of for-profit providers, where workers and service users are given the autonomy to organise to meet their needs.[28] Against the energy inefficiency and labour-intensive character of the single-family household, Hester has noted that a 'renewed emphasis on community resources, collective housing, and socialised care practices could offer real opportunities for restructuring social reproduction'.[29] Alongside this, co-production and democratic ownership of services with care workers and for people living with disabilities should be supported, ensuring everyone is equipped with the resources and capability to live as autonomously as desired.

Our healthcare systems have been stretched to breaking point by a decade of austerity and the traumatic demands of coronavirus. Environmental breakdown will only worsen health outcomes, with poor and marginalised communities bearing the brunt of environmental racism. Globally, the effects of environmental breakdown are likely to wipe out recent gains in public health, falling hardest on countries deprived of the chance to build up their healthcare capacity.[30] Public health interventions must embed equity, addressing racialised and class-linked environmental harms. Healthcare should be organised everywhere as a human right and a public good, given the investment it needs. And caring work should conjoin environmental and social reproduction: a world where urban rewilders lacing non-human life into the fabric of our towns and cities, and construction workers building a low-carbon future, are both forms of sustainable care work. Greening work is about more than economic restructuring – it is about reimagining how and where we live, our public spaces and homes, as spaces of everyday beauty and communal luxury.

To scale an alternative future of work, industrial strategies for the everyday economy are required. These must nurture the foundational economy: the non-tradeable, non-exportable sector of the economy 'that provides goods and services taken for granted by all members of the population and is therefore territorially distributed'.[31] As the Centre for Research on Socio-Culture Change's 2013 'Manifesto for the Foundational Economy' argued, these sectors operate with a form of 'social franchise', whether 'because it is directly or de-facto franchised by the state, or because household spending and tax revenue sustains its activities which are therefore sheltered'.[32] Public institutions should use their immense leverage via this social licence to end insecure, low-wage work and embed just working and environmental practices through progressive procurement policies and standard setting. Critically, the 'return of the state' (again) during the coronavirus crisis also underscored the extent of the neoliberal erosion of the state's capacity through outsourcing, austerity, and the whittling down of institutional knowledge. Implementing successful industrial strategies will require a more ambitious, directional state – but also one developing new epistemologies, capacities and planning powers than under neoliberalism.

The reimagining of the household and practices of care should cojoin struggles against an extractive economy and structures of gendered oppression, opening the door to a world of more fulfilling care and expanded, radical kinship.[33]

Socialising the Future

'Free market' capitalism is intensely planned, a vast legal, financial and managerial apparatus directing economic development and organising work. Yet the power to plan is narrowly held. Controlling the majority of investment flows, the terms of the future are defined by how investors allocate capital, their concentrated decision-making power shaping the terms and direction of economic development – what is produced and how. Currently, it is directing us towards catastrophic environmental breakdown. Within the command and control economy of the firm, workers are typically subordinated with limited capacity to shape the nature and direction of their work.

If planning is ubiquitous but hierarchical, undemocratic, and generating social and environmental harms, the challenge is to democratise it, organising economic life by social control not private power. Existing forms of workplace management, economic planning, and technological development too often attenuate human creativity and dismantle solidarity; we need to radically reimagine those same tools to serve people and planet.[34]

Anchoring efforts to restructure the economy in the direction of environmental and economic justice should be a set of interlocking 'green' industrial strategies. These should use the tools of economic coordination, planning, public investment and democratic ownership to scale sectors and technologies foundational to a sustainable future. Socially embedded markets are an important mechanism of change. They can be spaces of mutual recognition and interdependence, if organised on an equitable basis. But on their own, and in the dominant, unequal and financialised capitalist market forms we have today, markets will struggle to drive the reorganisation of production and consumption at the pace and scale required. They are blind to, and can amplify, myriad injustices, and fail to account for environmental externalities that threaten our collective futures.[35] Ambitious industrial strategies can extend social, democratic control over the terms and direction of production through new forms of economic coordination. And they can prioritise good work for all those who want it: well-rewarded, secure, unionised, rewarding and purposeful.[36]

The managed acceleration of automation should be at the core of a green industrial strategy. The spectre of mass unemployment has been raised – as both a threat and a promise.[37] But rather than eliminate work, the productivity gains generated by the integration of new technologies are likely to be recirculated, producing new sources of wealth, reshaping whole sectors and transforming the type of tasks humans perform. Managed poorly – absent democratic direction – automation could create a 'paradox of plenty': societies would be far richer in aggregate, but, for many individuals and communities, technological change would reinforce inequalities of power and reward. Managed well, it could provide the conditions for shared plenty. Automation is thus a question of distribution, not just production.[38]

Without countervailing power, automating technologies are likely to deepen existing inequalities. Nevertheless, if we can democratise technologies – and socialise the capital underpinning their development and use – automation could underpin an economy where technological change endows people with more resources and power, enabling 'deep freedom': the 'empowerment of the ordinary person – a raising up of ordinary life to a higher plane of intensity, scope and capability'.[39] That will require the promotion of pro-labour technologies in the 'everyday economy' and household, not just in frontier firms.

A new institutional arrangement for the democratic control of digital, data and knowledge infrastructures is required to reimagine work. The socialist calculation debate of the interwar period – centred on how to perform economic calculations absent market valuations – risks being answered today, not in favour of deepening real freedom, but on the extractive terms of the digital oligarchs. Universal platforms, ubiquitous surveillance apparatuses, ever-more granular forms of data, and vast, growing computational power are together creating the conditions for quantifying 'value' beyond market price: from individual credit ratings to the feedback infrastructures of vast corporations which can signal desires and needs without price and at speed, in ways early neoliberals denied was possible. They have increasingly perfected capitalist rationality,[40] and are set to emerge from the coronavirus lockdown more powerful than ever. This growing power – with the ambition to render all life calculable – can extend financialisation, intimate forms of control, and wealth extraction to unprecedented degrees. Or, alternatively, we can begin experimenting at scale with participative economics, cybernetic planning, distributed networking and digital feedback loops to improve working conditions, and support decentralised economic planning and production for non-market needs.[41]

The politicisation of technologies must be matched by the struggle for twenty-first-century workplace democracy: the democratisation of planning and the extension of social control over decision-making at work. Workers, waged and unwaged, are subject to unfreedom. The absence of autonomy relates not only to the individual's sense of their ability to act; it also lies in barriers to the collective pursuit of the transformation of joint activity. In a world of work transformed, workers should be better equipped to act together. A just, sustainable world of work must embed institutions for genuine economic democracy, giving workers and other stakeholders real power to meaningfully influence their work and workplaces. Traditional approaches such as co-determination and works councils, as is common in continental Europe, can lay the foundations for the democratic negotiation of work. Collective bargaining at a sectoral and enterprise level – and potentially even a global level – should be rolled out across economies and trade union rights should be institutionalised as basic entitlements. Capital's ability to coordinate should be matched by labour's ability to cohere and bargain over the terms of the future. This extends to challenging the injustices that weave through the formal and informal supply chains of the global economy, which must be central to any programme to transform work outside of formal employment in the Global North.[42] In an era of secular stagnation, and the disruptions and limitations that environmental crisis imposes, any reliance on the balm of high GDP growth is over; in a low-growth world, the struggle over the surplus will be more critical than ever. Workers everywhere must be re-equipped with the legal and institutional means to claim their share and have their income supplemented by a rising social wage. But we must go further.

Work in the twenty-first century should be organised by a new set of institutions and practices, including multi-stakeholder boards, general assemblies, participatory planning processes, heightened transparency and accountability standards, and the use of experimental participative tools such as commons-based peer-production and fabrication techniques to shape the development of work and technology.[43]

Deepening democratic practices and participation is the route to better working conditions and a more rapid transition to sustainability.[44] Current techniques for governing work and the workplace disguise or render impossible worker management – and marginalises the tacit knowledge of workers. This creates the appearance, though not actuality, of depoliticisation.[45] On the other hand, by better incorporating the practical knowledge and capacity of ordinary workers, participative economic democracy can help both democratise and decarbonise work. People on the front line can design and implement the transition. This must push back against efforts to deskill work and obscure awareness.[46] Think, for example, of an Uber driver not knowing what happens within the app, or the developers of the app itself only working on one modular element. Yet supported by institutions that take seriously the capacity of all workers to be collective agents of transformation, the extraordinary practical knowledge of ordinary people can be used to reimagine work – and our environments – towards justice and sustainability.

Scaling Public Affluence

In place of the economics of the twentieth century, rooted in the distribution of income through waged labour, we need a twenty-first-century vision where the resources and capabilities needed to flourish are provided through an expanded public realm and universal economic rights.[47] This must institutionalise a fundamental claim that should underpin ecosocialism: 'Abundance is not a technological threshold but a social relationship, undergirded by the principle that one's means of existence will not be at stake in any of one's relationships.'[48] The heart of politics is the struggle to democratise that relationship.

We must decommodify if we are to thrive.[49] Universal provision of what we need to live freely, securely and well can form the foundations of a more secure, just society. From an expansion of beautifully designed public housing to free education; from a new era of public parks and leisure facilities to guaranteeing access to health, care and twenty-first-century communications networks as so many rights, the scaling up of the universal basic services agenda holds transformative promise.[50] Socialising a growing share of the cost of living through the scaling of public affluence is a vital step beyond the logics of capitalism: we should all be able to live well beyond the market.

Economic and environmental justice is also best served by ambitious decommodification: pooling resources, shrinking environmental-intensive private consumption, and reducing waste,[51] with expanded public services providing the basis for a higher quality of life for everyone on a lower material footprint.[52] The rallying cry of an agenda for a time of breakdown is clear: 'Public luxury, private sufficiency'.[53]

In an age of environmental crisis and stark inequality, scaling public affluence expands our collective freedom and progressively shrinks the realm of society organised by the imperatives of profit maximisation. If austerity and privatisation – the transfer of social wealth to private hands – has pushed the burden of social reproduction onto private households and individuals,[54] the Universal Basic Services agenda can challenge this by socialising much of its costs. By guaranteeing access to well-designed services and resources, centred on meeting needs, we can ensure that everyone can live a secure, full life, pursue their chosen ends, and participate in their community.[55] Decommodifying life's necessities is the foundational step beyond the realm of necessity towards that of deep freedom. By removing from market exchange the goods and services we need to live well, the pressure on individuals and households to enter into wage relations is radically reduced, destressing the link between the need for income, waged labour, and the growth of production and consumption. And the ability to live well outside of the market can rebalance power between employers and workers, providing a critical lever to improve working conditions by giving everyone the power to say no to bad terms and work.

Decommodification is also critical to challenging rentier power. Growth no longer substantially benefits the majority, instead flowing predominantly to asset holders, rent-seekers and the high-paid: 'growth-based economies are more and more exposed as fundamentally predatory and extractive systems, with financialisation (of housing, utilities, transport, basic services) acting as a key mechanism to extract revenue from everyone else, and concentrate wealth at the top.'[56] The provision of goods and services we all rely on free at the point of use eliminates a vital source of rent extraction.

Guaranteeing Security

One further, much-discussed idea to bolster freedom outside the market is a universal basic income (UBI), an unconditional stipend provided to all residents within a particular geography or population.[57] The economic effects of coronavirus, by highlighting our co-dependency and vulnerability, have underscored and normalised the case for a guaranteed income for everyone. A progressively designed UBI would undermine relations of domination and go some way to redistributing resources, though arguably less than more targeted redistributive measures. Powerful and positive though its effect could be, it should be seen as a complement, not a panacea. Many of the bolder claims made on its behalf fail to adequately acknowledge – or address – the challenges and opportunity costs a transformative UBI would involve. Should our priority in an age of environmental breakdown really be individual cash transfers, rather than strengthening collective, solidaristic institutions that attend to common needs? And, politically, 'a universal basic income high enough to be genuinely liberating would require enormous expropriation of businesses and wealthy people. Consequently, there is no chance of its passage until there is an organized working class already powerful enough to extract it.'[58] But that level of power – and the economic transformation it implies – would most likely largely obviate the need for a UBI in such a context. Cohering labour power is a more pressing challenge, beginning in the nodal points of global supply chains where labour retains great, under-recognised, leverage. Without that transformation, a UBI defined by the Right risks being a wedge for deeper marketisation and the stripping back of public provision.

Given this, a universal basic income is a potent rallying

cry for an economy that guarantees security for all. In the current context, though, place-based experimentation and trials that can be scaled up are perhaps a more useful starting point rather than relying on UBI as the main tool to effect deep transformation. A more immediate goal in the struggle to expand freedom is the transformation of our dehumanising, managerial welfare systems into sources of genuine security and empowerment. Building on the job retention schemes and emergency welfare provision that emerged in response to coronavirus, a comprehensive, generous, non-conditional, non-means-tested at the point of access, minimum income floor should be introduced to radically improve the welfare and security of all.[59] This would not just redistribute resources; it would redistribute power, giving people a greater ability to negotiate the forms and quantity of work they want, out of choice, not necessity.

Reclaiming Time

It is not just the type of work we do that must change to address the entwined environmental and inequality crisis. We must also change the amount that we work. Over its history, capitalism has consistently transformed how we measure and organise time.[60] From the time discipline of the traditional factory, to the Amazon wristband that invasively tracks how warehouse workers use their time, the invention of capitalist time has prompted the struggle for a real-world utopia: a world beyond work. That same ambition, of the redistribution of time from the command of capital towards self-direction and the pursuit of freely chosen ends, must anchor the move from an extractive economy to substantively democratic societies and economies of greater collective and individual autonomy. We must reclaim the most finite resource of all: our time.[61]

There are several methods we could use to achieve this without loss of pay: making the progressive shortening of the working week part of collective bargaining, or gradual increases in paid holiday. Reclaiming time will require concrete steps and careful management, attentive to the needs of the most vulnerable such as caregivers and the insecurely employed; a four-day week is an emancipatory goal to aim for but does not suit the needs of all workers, especially those who are marginalised and insecure. The key is expanding leisure time for all. Of course, in and of itself, the progressive shortening of the working week and the expansion of paid holiday would not dismantle extractivism. All the same, as a 'non-reformist reform' it is 'a tangible and achievable shift in the equilibrium of power and an improvement on the everyday lives of the population'.[62]

As time spent working declines, aided by new labour-saving technologies, the capacity to play, create and experiment should expand. Environmentally abundant leisure spaces – parks and theatres, galleries and community spaces, sport centres, repair and fabrication workshops, communal kitchens and gardens, and so on – that support culture, creativity and play in all their dimensions should be prioritised in place of carbon-intensive production and consumption.[63]

Leisure under neoliberalism is increasingly extractive and commodified, from our social media data to the ways we are encouraged to treat our own personalities as maximisable. But with leisure reclaimed, we could see a transformation of leisure into something entirely new. We have a whole world of pleasure to win.

Choosing Life

A collective project of world-making is required to overcome the crises of extractive capitalism. That will require deploying the tools neoliberalism has long sought to neuter: strengthened labour power, through the radical extension of sectoral collective bargaining and the re-regulation of labour markets; social reproduction put at the heart of the politics of work; strategies to shape technologies to meet social needs; and the democratisation of work through a new ecology of planning and control that redistributes power and resources. In so doing, the transformation of work and leisure through deep decommodification and the democratisation of economic power can help us all live better. We want neighbourhoods that bustle with life, public housing and spaces of everyday luxury attentive to human needs; we aspire to fulfilling work that takes seriously the dignity, worth and extraordinary capability contained in all of us.

9
A POLITICS FOR LIFE

Ecology is permanent economy.

Sunderlal Bahuguna

We don't have time to sit on our hands as our planet burns.
For young people, climate change is bigger than election or
re-election. It's life or death.

Alexandria Ocasio-Cortez

There are many ways to make change. Too often, the 'great men' view of history obscures how change often results from the concerted efforts and struggles of all people, who have learned, organised, fought and won. Yet, in the age of environmental breakdown, and with the extraordinary reconfigurations unleashed by Covid-19, it is more important than ever for us to understand not just how change can come from the 'top down' and the 'bottom up', but how this distinction can be abolished altogether.

Two stories may be helpful here. The first happened in the 1970s, in a valley in the rural Himalayas, and offers an important lesson in how to fight together and win.

Life-Making vs Extractivism

Vast forests blanketed the mountains that dominate the region of Uttarakhand in northern India. For centuries, local people had coexisted with the forests, foraging fruits, growing fodder for cattle, and using wood to make houses. In turn, the forests purified the water running through the ground, their roots stabilising the soil, holding it in place throughout the rainy season. During the Sino-Indian war of 1962 the Indian army built a network of roads, connecting the region to the wider world. Along these roads came companies who sought to exploit the forests for profit. These companies enjoyed the patronage of India's restrictive forestry policy, an inheritance from British imperial rule, and took advantage to strip back the forest, denuding vast tracts of land. In response, the local people organised, creating workers' cooperatives that won forest auction rights and used the land to collect and sell roots and herbs and to manufacture turpentine and resin from pine sap. But they struggled as the large logging companies used their money and influence to gain a greater share of the forests. Soon, deforestation began to rob the region of its protection against monsoon rains and in 1970 a major flood killed hundreds, as mountainous slopes subsided, stripped bare of the trees that once held the land in place. It is against this backdrop that the local people began to hug the trees.

In 1973, a workers' cooperative sought a licence for a small plot of land to use trees to manufacture farm tools. The licence was denied and, instead, a large sports company was granted rights over a larger plot. On the day the company was to begin felling, local people arrived and embraced the trees, blocking the path of the loggers. Astonished, the company withdrew and the forestry authorities demanded the locals do the same. But they refused, and after doggedly holding out, were granted their licence while the sports company's permit was revoked. The next major confrontation came in 1974 in the village of Reni, the site of the floods in 1970, where over 2,000 trees were to be felled. Facing down the threats of armed loggers, a group of local women stepped up to hug the trees, sparking a four-day standoff. The loggers left, defeated by bold, non-violent action. In the wake of this moment, the Chipko Movement was born, named after the Hindi word for clinging or hugging. Before long, the Indian government agreed to ban commercial logging in the valley for a decade. Beyond this stunning victory, the movement was remarkable for the prominence of local women at the forefront of the campaign, and stressed the mutual dependence between humans and nature; its struggle lives on in many forms. In the words of Sunderlal Bahuguna, a leading member of the movement, 'ecology is permanent economy.'[1]

Meanwhile, on the other side of the world, David Koch

was running to be vice president of the United States of America.[2] David and his brother Charles owned Koch Industries, a sprawling multinational conglomerate of business interests, ranging from oil to financial trading, inherited from their father. The brothers resented government, seeing its web of laws and institutions as antithetical to individual liberty and a barrier to their personal enrichment. This led David to join the US Libertarian Party in 1980 and run for vice president, bankrolling a campaign that sought to abolish Social Security, gun control, financial regulation and the Department of Energy, reducing the function of government to the 'protection of individual rights'. David and his presidential running mate lost, scoring 1.1 per cent of the vote; but in Ronald Reagan's victory, wider ideological success was soon to be achieved.

In losing an election, the Koch brothers had found a surer strategy for realising a more libertarian, more profitable world. The 'nasty, corrupting business' of frontline politics had taught them that politicians were merely, in their words, 'actors playing out a script'. The game was to write the script. Instead of running for office again, they began pumping their burgeoning wealth into a web of organisations and politicians that argued for smaller government, less regulation and more profits for big businesses like theirs. In the 1990s, the Koch brothers turned this strategy on the biggest threat to their interests to date: the growing commitment of governments to act on environmental breakdown.

In 1991, the Koch-funded Cato Institute, an anti-government think tank, organised a conference called 'Global Environmental Crisis: Science or Politics?', which brought together the people and arguments that became hallmarks of a concerted Koch strategy to libel science and halt action to avert catastrophe. This strategy was soon being implemented by a pervasive, well-funded ecosystem of think tanks, media organisations, academic departments, 'grassroots' campaigns, legislative programmes and educational materials. As the planet warmed, the Koch brothers' net worth exploded. Though initially they opposed Trump, his election marked the latest in a long line of victories for them. In the words of Jane Mayer, a leading chronicler of the Kochs: 'Whether announcing his intention to withdraw from the Paris climate accord, placing shills from the oil and coal industries at the head of federal energy and environmental departments, or slashing taxes on corporations and the ultra-wealthy, Trump has delivered for the Kochs.' In 2018, indeed, Charles Koch announced to fellow political donors that 'we've made more progress in the last five years than I had in the previous fifty.'

In many ways, the five centuries leading up to our current moment, the terminal phase of environmental breakdown, can be characterised by the battle between the Chipko Movement and the Koch brothers; between everyday people – many of them in the Global South – resisting destruction and realising different models of living, and those elites who would see the world burn merely to maximise profits. In short, it is a history of life-making versus extractive suicide.[3]

And life-making has been losing. In response, we need both a Chipko and a Koch strategy. Both are needed to unleash the latent energy of communities across the world and ensure that the power of markets and governments, of the commons and a reimagined household economy, are realigned to support a great global effort to respond to environmental breakdown. Only with a top-down and a bottom-up strategy can we make a world in which those distinctions begin to blur, ending the chokehold of vested interests, giving everyone a stake in making change, and protecting our common humanity as destabilisation grows.

Winning Life

Both the Chipko Movement and the insider strategy of the Koch brothers were able to make changes in politics, and by extension, the environment. Their experience helps us answer the most urgent question facing us: how do we win so that life can thrive from out of the ruins? The environmental crisis is fundamentally a crisis of politics. We can therefore address it democratically and justly. We have the capability, ingenuity and resources to build sustainable, equitable economies, realise societies of shared plenty, and repair the natural systems ravaged by extractive capitalism. The challenge is how we mobilise to match the scale of the emergency and overcome the entrenched interests and popular inertia inhibiting adequate action. In recent years, the permafrost of 'there is no alternative' has cracked – we now need to break it wide open.

If the crisis is political, politics in all its dimensions is the best hope to rescue our future, renewing hope against the deadening claims of the past. Yet therein lies a problem. Social democratic parties are in long-term decline, struggling to navigate the shifting terrain of cultural, demographic, economic and technological forces reshaping societies, their historic coalitions disintegrating. And though movements and parties articulating a transformative agenda are surging everywhere, they are governing in too few places. Worse still, their opponents are winning. At best, status quo neoliberals, presiding over a technocratic centre, will enact a politics of anti-democratic managerialism that pursues sustainability, but too slowly and at the likely expense of justice.

The feedback loops of environmental breakdown do not automatically lead to a political realignment that favours coalitions committed to serious action. Growing destabilisation could see denialist conservatives morph into eco-ethnonationalists, recklessly defending the commanding heights of the extractive economy, an aggressive acceleration of inequality, and the increasingly violent policing of a world fragmented by environmental crisis. It offers a brutal common sense in response to an era of increasing disruption, one that taps into a cultural war played by authoritarian populists. Crisis can and will be seized by the far right to remake our politics if we let them. To avoid that fate, we need a clear plan to win, building on recent periods of advance, learning from missteps, and preparing the ground for a popular front capable of

effecting transformation. There are many tributaries feeding into that, but five steps must urgently be taken in the
terminal juncture of the 2020s.

First, we need a narrative. Building a different future
requires an alternative vision that is at once radical and
credible, a common sense for new times. This story should
be anchored in the gains ordinary people and communities
can win for themselves through a collective politics of
transformation. All wealthy industrialised economies must
limit environmental destruction as rapidly and justly as
possible: this is about building flourishing communities of
security, leisure and solidarity; green work that is rewarding and meaningful; and new forms of public luxury that
make everyday life more pleasant. An ambitious response
to crisis is a common project to help us thrive, not just
survive, not just making today's economy environmentally sustainable but building the democratic economy of
tomorrow. It seeks to protect and nourish simple desires:
common care, mutual solidarity, and the collective task of
keeping the world alive and flourishing.[4] The popular
response to Covid-19, of mutual aid, new forms of solidarity and the recognition of care as foundational work in
society, provides fertile ground to grow such a politics.
This is a project of modernity, owning the future and its
possibilities, while remaining anchored in a compelling
moral story and ethical appeal: the nurturing of life. The
true extremists, by contrast, are those who defend the
ramparts of the contemporary economy, content with a
smallness of ambition that guarantees deepening planetary

emergency. In telling this story, new voices and forms of leadership will be required, centring different experiences. Traversing the current conjuncture will also require a nimble politics attuned to the cultural as well as economic terrain, uniting interconnected struggles, communities and coalitions in aid of a politics fighting for life. It will need commitment to deep constitutional reform, democratising political power and experimenting with new forms of voice and participation, and decentralised governance. But it cannot rest on narrow electoralism. There is no single path or agent that can get us there: it must combine social movements and electoral projects committed to winning and transforming state power, party politics and that of organised workplaces, grassroots capacity-building and powerful campaigning.

Second, we should embrace constructive antagonism. Building an environmentally sustainable future should put all of us first. But the transition will not be painless or without its temporary losers. Where there are communities impacted by structural alterations to environmentally damaging employment, it is vital to develop credible plans for a just transition of those sectors. Ultimately, this must be led by people on the front line of change: workers, trade unions and those whose livelihoods are threatened by a poorly managed transition. Opposition from the entrenched elites of extractive capitalism is inevitable. If they stand opposed to the politics of life, their opposition must be met not with timidity but a clear and credible plan for how their power will be challenged, isolated and dismantled. And education remains supremely important; never forget that so many people do not know how bad it has got and why this has happened. But don't think that education on its own will drive progressive action. Change is not that easy.

Third, it must have a clear sense of the coalition that can build and win state power – and how the state can be transformed and democratised to deliver environmental and social justice. This should be a majoritarian alliance united by the fight for a liveable future, particularly those on the front line of transformation. But that requires centring the needs and voices of today's working people, not a nostalgic, masculinised imaginary. The new working class of the twenty-first century is more likely to be employed in care work than heavy industry. It is multi-ethnic, diverse and drawn from migrant communities. Younger generations and many working-age people generally who have suffered under neoliberalism, with decades of poor pay, insecure housing and personal debt, are now set to inherit a planet on fire, and it is they who are therefore the central elements of a popular bloc of working people willing and able to respond to crisis. Covid-19 bookmarks a terrible economic decade for the young; politics must renew hope or face the backlash. We will need a more ambitious and inclusive intergenerational politics.[5] Without winning older generations to the politics of life, the political economy of extractivism will not be dismantled in time, if at all. To do so, we must embrace continuities with past projects of democratic renewal and inclusive civic pride, of struggles for social and economic emancipation, and show how they can lead to the security and dignity a new settlement can deliver. This coalition must cohere to contest and win state power, while being connected to and driven forward by social movements and their demands, in and against the state.

Fourth, we must prefigure the systemic change to come. A persuasive narrative is not enough. To fight corrosive scepticism about the ability of collective politics to achieve positive change, a scepticism deliberately fostered, wider systemic change must be prefigured in the here and now, with demonstrable proof of the credibility and success of alternative strategies for organising how we live and work. Capitalism does not organise all of life: vast swathes of production and distribution are organised within different, often more sustainable systems, from cooperatives to the commons to peer-to-peer production. A politics of place, through green community wealth building, can bring these alternatives to life, and challenge injustices where they exist. This requires a new chapter of ambitious municipalism, with towns and cities vital islands of experimentation, living the future by reimagining transport, care work and economic purpose today. Green enterprise is vital here, too. Entrepreneurialism by its nature breaks with convention and seeks to disrupt and invent; that includes breaking from the institutional arrangements of extractive capitalism, creating new forms of enterprise that are democratic and sustainable by design. But that requires scaling generative, not predatory, forms of enterprise. Prefiguring transformation also requires ideological contestation, building up the case for change through a wave of political education, developing an ecosystem of ideas generation and diffusion – from the shop floor, through think tanks, to new media spaces – that can make the transformative credible, reassuring and commonsensical.

Finally, strategic timing is essential. In an era where economics has disenchanted politics[6], it is hard to imagine realising our entire agenda all together and at once. This is particularly the case if people do not believe in the ability of politics to transform. So, we need a clear strategy of prioritisation. This requires targeting structural reforms that, instead of stabilising the status quo, make permanent changes to the social alignment of power, opening the possibility of compounding institutional divergence over time. From the expansion of paid holidays to transforming the rules governing the company, from instituting a new public banking ecosystem to scaling democratic media, the immediate implementation of key policies is crucial to the strategic success of any transformative governing project – building power and securing allies. The deployment of a green investment surge that can at once restructure and revive our economies in the wake of Covid-19 is urgent in this context. Moreover, our timing must be wise to the fact that destabilisation will grow, which risks strengthening the hand of ethnonationalism. Our strategies must be robust against this. Fighting against 2°C in a 1.8°C world will be vastly different to fighting against 1.5°C in a 1.2°C world, as we are now. Fighting against 2.5°C at 2°C, more so.

How We Can Win

The scale of change required may appear daunting, yet we know from our recent past that rapid transformation is possible, and for that we can thank Margaret Thatcher. Under her leadership, the Conservative Party transformed the British economy. Within a decade, the power of organised labour was dismantled, finance unleashed, and property relations transformed. Change was driven by a political project with a clear narrative and sharply framed enemies, acute strategic timing and organised class power, and effective prefiguration that combined to transformative effect.

Thatcherism was, in Stuart Hall's memorable phrase, a project of 'regressive modernisation'.[7] At once a political narrative of 'authoritarian populism' and neo-Victorian moralism, of the household budgeting of a greengrocer's daughter in Grantham translated to the economics of national accounting, it also presented a clear analysis of its times and a story of the future: only the New Right could modernise the UK, decisively ending the crisis of social democracy to build a property-owning democracy and restore national pride. The narrative groaned with antagonists, made of 'enemies within' and opponents abroad: the trade unions, 'Loony Left' local councils, over-mighty Brussels, and the 'wets' within her own party. There was a clear strategic ordering of action, a biding of time and building of strength. The Conservative manifesto of 1979, for example, contained only a small number of privatisations relative to those later in the decade, when Thatcher's position was more secure. Perhaps most famously, following the trajectory set out by the Ridley Plan – a 1977 report by Conservative MP Nicholas Ridley on how the next Conservative government could provoke, confront and defeat a major strike in a nationalised industry – Thatcher did not immediately challenge the National Union of Mineworkers (NUM), but rather began stockpiling coal and passing legislation, amid other measures that would later prove pivotal in defeating the strike. When the confrontation eventually came, her government was able to emerge triumphant, decisively breaking the power of the NUM in the epochal miners' strike of 1984–85. The path was cleared for the fragmentation of organised labour and the consolidation of Thatcherism.

This was an extreme approach; it should be remembered how much of an outlier the UK was in this period, with 40 per cent of the value of all privatisations among the OECD between 1979 and 1996 taking place in the UK alone.[8] These transformations rapidly built a successful electoral coalition, despite the devastating effects of Thatcher's economic policies on many parts of the country. The privatisation of public wealth through the sell-off of public utilities and housing helped cement a cross-class coalition by distributing material benefits and power to its recipients. The 'Big Bang' financial deregulation of the City of London in 1986 created a surging credit boom that enriched key elements of the Conservative coalition. Many of the most iconic policies of Thatcher's administration

– the Right to Buy one's council house, for instance – had been trialled in Conservative-run local councils prior to 1979. A dense web of think tanks and public intellectuals provided a sense of energy and ideological ballast to the project, from monetarist voices in academia to organisations like the Institute for Economic Affairs and Centre for Policy Studies, which generated a wave of ideas and policies for the Thatcherite wing of the Conservative Party. A smooth path ran from ideological prefiguring to local testing to national implementation. All the while, organised power in the form of deepening financialisation drove the project on.

The result: a country transformed, a new institutional configuration embedded at the heart of British capitalism, and enduring political domination.

Manifesto for a Planet on Fire

We live in the gathering ruins of extractive capitalism. Eco-ethnonationalism is on the march. The brutal shockwaves of Covid-19 have cast into question the fundamentals of our economy and ways of living; with breakdown accelerating, there will be no return to 'normality'. Going forward, our response will need to be deeper, more systemic, and better able to heal the accumulated harms of the status quo. Ours must be an ambition greater even than Thatcher's, albeit employed for radically different means and ends; this late hour necessitates it. This is a manifesto for a planet on fire, ten steps to build societies of flourishing and meaning.

1. A new purpose

Our goal is a deep and purposeful reorganisation of our
economy so that it is democratic, sustainable and equal by
design. That goal will require rewiring the direction and
purpose of economic activity. The pursuit of GDP growth
as society's central mission must be consigned to the past. A
flawed statistical measurement that emerged in the 1930s as
part of efforts to construct the idea of a national economy, it
is inappropriate for guiding the rapidly changing economy
of the 2020s.[9] Growth will remain the goal, but growth of a
different kind, of social and environmental progress, of
equity and care. Public policy should be measured by how
it lowers environmental impacts to within sustainable limits
at the same time as improving the conditions for human
flourishing by expanding leisure time, equally sharing the
fruits of economic activity, increasing wellbeing and improving mental and physical health, boosting communal affluence and providing more meaningful forms of work. Of
course, there will be limits: destructive and unsustainable
economic activity must be rapidly curtailed and demand for
highly carbon-intensive activities managed downwards. We
cannot afford to use GDP growth for its own sake as the
lodestar guiding policy, but nor should we measure success
by blindly 'degrowing' the economy, which would take the
same flawed measurement of GDP to shape public policy
aims, but apply it in reverse. Instead, we must build an economy based on a new purpose: to imagine and grow new
forms of flourishing, abundance and sustainable wealth.

2. *Financing to flourish*

Under social control, the world-shaping power of the financial system can build a future of collective flourishing. Yet there needs to be a series of deep shifts in the organisation and purpose of the financial system to achieve this. First, central banking should actively steer our economies through rapid transition, directing credit towards green sectors and repressing environmentally destructive investment, based on democratically mandated goals.

Second, we must tame the power of private finance. That requires new rules to ensure it is an effective servant of society, not a bad master: greening collateral and capital requirements to penalise dirty investment and make green investment standard. Third, we need a new ecosystem of mission-oriented public banking that can derisk investment and fund the technologies, infrastructures and forms of enterprise we need, financing activities that serve social and environmental interests. Finally, rapid transformation depends on a step change in public investment, beginning with a green stimulus to recover and reimagine our economies post-Covid-19. In turn, that depends on fiscal rules fit for an age of breakdown and an expansive new relationship between central banks and national treasuries to unlock a transformative, affordable increase in debt-financed public investment. Taken together, democratic control of finance can resuscitate the future.

3. Owning the future

Extractive capitalism's deep institutional arrangements are a dynamo of crisis. Addressing our environmental and economic challenges therefore requires a deep overhaul of how the economy is conceived and a rewriting of the 'rules of the game'. Only by redesigning how our economy is owned, and to what purpose, can we centre new values, relationships and goals at its heart. We must have a strategy for scaling a new ecosystem of ownership and control that is democratic and sustainable by design. We must make enterprise into a generative, collective endeavour, putting power back into the hands of the many.

That should start by 'greening' the company. New rules are needed to turn it from a vehicle for maximising profits for the few towards an institution of the commons: a collective and democratic endeavour to create value sustainably and justly, balancing the interests of all key stakeholders, not just capital, but labour, society and, critically, the environment. The asset management industry's power, based on the control of other people's money and producing stark inequalities and harm for environment and society, can be tamed by a simple principle: ensuring the ultimate beneficiary can exercise their voting rights, with new rules to promote the investment of pension wealth in a sustainable future.

To truly transform the economy, we need to democratise capital, so that all may share in our common wealth. New institutions can facilitate this. Worker ownership funds and steps to scale cooperatives, employee-owned enterprise and the solidarity economy can give workers greater collective control and individual autonomy – and ensure they share in the success of their company. A wave of social wealth funds and People's Asset Manager–like institutions can help socialise capital at scale and ensure everyone has a stake and a say in the common wealth. And twenty-first-century forms of democratic public ownership can ensure that society's foundational institutions and services serve people, not profit.

In doing so, transforming wealth and control from the few to the many can build a powerful coalition behind change, redistributing not just material benefits, but economic and political power in the economy towards working people and their families. By embedding new purpose at the heart of enterprise, a fundamental institution in the shaping and organising of natural systems can be turned into an actor for sustainability and justice.

4. A twenty-first-century commons

The logic driving planetary emergency – enclosure, extraction, expansion – must be replaced by a new, twenty-first-century commons. From data, digital technologies and IP, to land and natural resources, to media and platforms, the infrastructures and institutions upon which we all depend should be managed for the common good. This will require rewiring our legal infrastructure to enable new forms of democratic association and governance over critical resources to scale. New models of land stewardship should support sustainable land use practices, including rewilding and regenerative agriculture, focused on repair, nurture and sustenance. New institutions like the Common Ground Trust, community land trusts and public–common partnerships can reclaim land from its original enclosure and underpin a new wave of environmentally sustainable, affordable and beautiful housing. A new network of data trusts can reclaim social data from the walled gardens of the tech monopolies, creating a more innovative and open digital economy, and enable societies to plan more effectively, from decarbonising transport systems to energy grids. Digital platforms should be transformed from profit-driven spaces into sites of cooperative management, while new media should be scaled. Strategies for the democratisation of technology more generally from a new commons-based IP regime to public ownership of vital digital infrastructures can ensure technical infrastructures are arranged in generative configurations that support environmental and social justice.[10]

By creating tangible material benefits and nurturing new forms of democratic engagement, a twenty-first-century commons can prefigure wider systems change. A thriving commons can channel social and economic forces in favour of pluralism, openness and creation, against the inequalities and exclusions generated by the economics of enclosure.

5. Living, not just surviving

If the twentieth century distributed income through waged labour, sustainable economies of the twenty-first century must steadily expand public affluence, giving everyone the means to live well through the provision of free public goods and services. 'Public luxury for all' should be the slogan of a post-carbon age. Achieving this requires an ambitious programme centred on decommodification and decarbonisation. Housing should be a right, not a commodity, supported by a new programme of public and cooperative housing. A wave of investment in the spaces and materials we need to thrive – from leafy parks to community theatres, from leisure centres to bustling marketplaces – can build new spaces for leisure and communing. Municipal urbanism can rescue our towns and cities from the grip of the motor-car, based on a new vision of multi-modal transport that is decarbonised and free at the point of use, building healthier towns, cities and rural communities. And access to the fundamentals of life – from digital connectivity and beautiful built environments to good-quality food and cooking facilities – can expand the ability of all to live well and freely.

A new agenda of decommodification plus environmental sustainability can build a broad coalition, anchored in a twenty-first century economic vision, providing an alternative for all those constrained by the limits that markets can impose and those who want a society built on shared experience, a common life, and deeper individual and collective freedom.

6. *Working for life*

Unless we transform work – its rhythms, purposes and organisation, anchored in new forms of solidarity, kinship and creativity – we cannot repair existing harms nor build the foundations for a just, free, emancipated post-carbon and environmentally sustainable future. A new politics of work must start by guaranteeing well-paid, rewarding work for all those who want it. But that will require building or strengthening institutions of twenty-first-century economic democracy that give working people collective power and individual autonomy: extending collective bargaining and works councils to give labour its fair share and voice, but also requiring new avenues for participation and control in everyday working lives. We cannot work with more autonomy, purpose and creativity without dissolving the authoritarian relationship at the heart of work – and rectifying deep inequalities of power within the economy. New forms of universal entitlement, from a minimum income guarantee to a universal basic income, can rebalance underlying asymmetries in resources and bargaining power. Ambitious industrial strategies – with working people and unions at the forefront – can restructure production and consumption, supporting just social and ecological reproduction and driving the mass deployment and adoption of environmentally sustainable technologies and infrastructures. The managed acceleration of automating technologies, shaped via new forms of regulation, ownership and investment, can help meet social and environmental needs rather than just profit maximisation.

Work isn't working for many. The coalition for contesting the hierarchies and imbalances of contemporary work is broad and deep; the challenge is to convoke a multi-class coalition, led by working people, that can organise and win.

7. Caring and playing

Our response to environmental breakdown must centre and value new ways of reproducing and caring for life. The green economy will be an economy rooted in the work of care, nurture and repair. How care is recognised, honoured and rewarded must shift, from industrial strategies for the 'foundational economy' to the commoning and socialising of care resources and practices, and the co-production of health services. Universal public childcare and decent adult social care are preconditions and absolutes. Along the way, a politics that stresses the inseparability of care, work and the environment can dismantle the gendered hierarchies and unequal distribution of time and effort that sustain social reproduction in today's economy. Reclaiming time for leisure by 'banking' the gains of technology through more paid leave and the managed reduction of the average working week without the loss of pay should be introduced as a crucial step towards sustainable societies with greater freedom, though it is a process that will require careful management to ensure fairness. Done well, a new era of vitality, play and mutuality is within reach.

8. *Energising our lives*

The logic and behaviour of the for-profit fossil fuel corporation – to transform nature into profit regardless of the environmental and social consequences – presents a direct threat to a liveable planet. We need to sever the link between fossil fuel energy and capitalism. Any serious strategy for a post-carbon economy must then have a plan for both the managed decline of fossil fuel production and a rapid scaling up of renewable sources of energy, built on ethical supply chains and new forms of ownership. Indeed, unless we can sever the knot between energy and fossil-capital, efforts elsewhere will not be enough.

To that end, no new infrastructures or projects for the extraction or transportation of fossil fuel should be built; new government permits for such activity must cease, as must the vast network of subsidies supporting fossil fuel production. Fair carbon taxes should be introduced to raise the cost of production and new rules should shift institutional investors out of carbon-intensive assets into green energy systems. And if this fails to trigger a rapid shift in behaviours and environmental impacts, the fossil fuel majors should be taken into public ownership to restructure their activity and drive a managed decline in fossil fuel production. That transition must be just; workers in impacted sectors and wider supply chains should be given support to transition into new forms of work. The opportunities arising from the coronavirus shock cannot be lost.

Scaling renewable energy generation will require a combination of explicit subsidies for green energy, clear signals to producers and consumers, and targeted industrial strategies to grow out the capability of the sector. Together, this can drive the rapid deployment of renewable technologies. New models of decentralised community ownership for the generation and distribution of power are needed too, as is public ownership of national grids. Taking advantage of the capacity of digitisation to decentralise and localise production, the wealth and opportunities of transition can be anchored in place. At the same time, a wave of investment is needed to roll out smart grids and heating systems, building an affordable, secure post-carbon future. The future is renewable power plus shared abundance.

9. Cooperating to win

Cooperation is presently at a low ebb, eroded by austerity, inequality and systemic oppression. Neoliberalism has tied the hands of the state at precisely the moment we need its unique capacity the most. Rebuilding public capacity for cooperation starts by endowing communities with more oversight, democratic input or outright control over economic and social institutions, from energy to education to health. This is already happening: if done right, municipalism and community wealth-building could drive rapid sustainability from the bottom up, giving everyone a local role to play in a global effort to respond to environmental breakdown. Locally decided rules can ensure investments are channelled into clean development and public money used to promote organisations that heal environmental destruction in the process of providing better public services. Green businesses and products and services can promote wellbeing and community over endless material consumption, while local currencies can lock in sustainable investment, with shared ownership models allowing local people to decide where investments are made. The community response to the coronavirus pandemic provides an organic model and an opportunity for change.

Local action should be connected and coordinated by a centralised and better democratised state that can draw on the full array of resources at its disposal, marshalling an all-society response, giving everyone a role as we rise to an all-encompassing emergency.

This will require states to develop their institutional capacity to connect communities, enabling accelerated action to reduce environmental impacts in a just way. Examples of this are all around; in the UK, the National Health Service leads the world in reducing its environmental impact, leadership which derives from the local efforts of its staff and the supporting role of its central buyers and organisers. Moreover, in providing high-quality healthcare, free at the point of use, public health systems show that there's always been an alternative to the conflicts and inequalities of neoliberalism, acting as a bolster against the shocks to come.

10. Ending empire

The history of environmental breakdown is the story of imperialism, of a world seared by colonialism and the successive waves of extractive expansion. Over the last four decades, a neoliberal model of economic globalisation has evolved from the end of formal empire, preserving and extending the inequalities and exploitation of the imperial era. In turn, people and countries have been pitted against one another, eroding international cooperation at precisely the wrong moment. Environmental breakdown is an international problem and so any credible agenda for global cooperation must be founded on a positive-sum internationalism and be capable of collectively managing the enormous risks of an environmentally destabilising world. Achieving this requires all countries to take part in telling an honest story of how the past is linked to a destabilising present. Wealthy nations bear unique responsibilities bequeathed by the benefits of five centuries of carbon largesse and imperial extraction. At this late stage, the UN mantra of 'common but differentiated responsibilities' means that each nation's contribution to mitigating environmental breakdown should be proportionate to the historical damage caused, not just present and future impact.

Contributing a 'fair share' goes beyond reducing the environmental damage. It also means acting to realise a more democratic and equitable global economy. Unequal financial and trade arrangements, holding back those least responsible for the breakdown, must be dismantled and replaced by a global system of investment and solidarity. Reform of multilateral institutions, such as the International Monetary Fund and the World Bank, should have them open their halls to excluded voices, prioritising the capacity of local people over multinationals to chart a new social and environmental course. The rule of the corporation should be placed below that of law, where it belongs, with international trade treaties no longer demoting human and environmental rights. And in a destabilising world, we can no longer afford to let nations disobey the rules of international law and sustainability and human rights commitments. The vicious spiral of illegal war and lopsided treaties becomes a self-fulfilling prophecy as seas rise, soils spoil, and ethnonationalists march. Instead, rights and protections must be given to those whose homes are lost and those for whom this is already or will become an existential threat. They must become world leaders, their struggle for justice and survival acting to remind everyone of how we got here – and how we get out of it. We simply cannot afford to marginalise whole swathes of the world if we hope to manage the compounding risks of the age of environmental breakdown.

Go Further, in Hope

The transition to an environmentally sustainable future will not be without difficult choices and trade-offs, nor will it create a world without pain. What we have outlined here are the contours of a strategy for the avoidance of unnecessary misery, for rescuing our futures, human and non-human alike. It gives us an opportunity to repair, as far as we are able, the crisis that has already arrived for so many people and places. Though no utopia, it can bring about sustainable abundance, anchored in a new set of values, relations, practices and institutions. Over time, these may seem as remote from the logics of our extractive present as industrial capitalism was from feudalism, or Fordist mass production from today's digitalism: a world of communal luxury and generative enterprise, of shared care and the nourishing of life. It is a project already underway, borne aloft by Green New Dealers, transition towns, indigenous communities, school strikers, permaculturalists, progressive politicians, Global South diplomats – each day, the list grows ever longer. You are in good company.

To salvage the future will require organisation, courage and, bluntly, luck; there are no guarantees. Life now inescapably exists under the growing shadow of catastrophe. We must accept that we will not 'solve' environmental breakdown. The level of destruction and the inertia in natural systems means things will get worse, potentially much worse, before extreme destabilisation can be halted and harms repaired. To win an alternative future means recognising the world is not static. Almost everything will change, and change utterly, often in ways that militate against our agenda. Science has provided us with a unique window into the future and unless our strategies are wise to growing destabilisation, we risk being overwhelmed by compounding crises. This means pre-empting those trends which favour ethnonationalism and acting now, developing and executing a medium-term strategy that's wise to the realities of growing destabilisation. We need a bolder, more inclusive migration narrative, based on humane and fair movement, to draw the poison of the nativist story and prepare for a world with more people on the move. We need to think about war and conflict and nuclear disarmament, opening up discussions of foreign and defence policy as the conditions for peace shrink. We need to mobilise a wide constituency of political backgrounds, both at home and across the world – young and old, workers and the retired, renters and asset owners, north and south – as only together can we overcome our greatest crisis. For we are the resistance to spiralling collapse, to the ethnonationalist barbarity of walls and wars.

A unique burden is placed on millennials and those even younger. These generations, at least in the West, are not just the first to 'have it worse than their parents'; they will inherit the catastrophic legacy of extractive capitalism and environmental breakdown. So, if you are of this age, face the future, and the grief and fear that lies therein. Then reach out, meet others, organise, hold onto one another, and never allow anyone to take from you the one unshakable truth upon which we will build a new world: no matter how bad it gets, an alternative is possible and will always be possible. It is easy to slip into despair, to conclude that 'it's too late' and that some Hollywood-style apocalypse lurks around the corner. It is natural to feel this way but to think it is to profoundly misread this moment. As Eric Holthaus has written, 'You are alive at just the right moment to change everything.'[11] We have moved to the next phase of our fight against the environmental crisis, a state of planetary emergency in which we must contend with global destabilisation while stewarding an unprecedented socioeconomic transition. But this is no 'paradise lost'; if you thought all was well at 350ppm or 1°C, you weren't paying attention. If you're as yet relatively sheltered from the storms and shocks or you're older and find yourself overwhelmed by the spiralling crisis, please do not slip into despair. Your role is to rejoin the fight, supporting those on the very front line, those on low incomes, people of colour, the young, those in the Global South. Boomers aren't allowed to be doomers. How the world reacts to 1.5°C, 2°C, 3°C or even 4°C is, in so many ways, down to politics. It is in the dark recesses of fear and pain, where the status quo offers little hope, that ethnonationalism grows, dictating that climate refugees should be left to drown and walls be built. It is a politics that is the enemy of life. Instead, we tell brave, honest narratives of the destruction and injustice, stepping up to embrace the world with hope and possibility as catastrophe unfolds, clinging on to save those around us while we fight with all our might to calm the storms. In this, the terminal phase of a centuries-old global struggle for dignity and life, so much is and must be possible.

Right now, we are losing the struggle for life. Recovering

our future will require supreme imagination, care and collaboration, a common daring to thrive, not just to survive. Together we can escape the ruins, charting a new way forward: a sustainable future anchored in democracy, justice and mutual solidarity, in a world fit for life, in all its finitude and wonder.

ACKNOWLEDGMENTS

We would like to thank our colleagues at IPPR, Common Wealth, the Economic Change Unit and the New Economics Foundation. We are indebted to those whose books inspired this one, including Raj Patel, Jason Moore, Simon Lewis, Mark Maslin, Daniel Aldana Cohen, Alyssa Battistoni, Thea Riofrancos, Kate Aronoff, Geoff Mann, Jedediah Britton-Purdy, Joel Wainwright, Quinn Slobodian, Kate Raworth and many, many others. Most particular thanks go to the fantastic team at Verso for making it possible, and to John Merrick, our editor and friend, whose energy and patience we have appreciated through every moment. And we thank each other, particularly for all the companionship and the scheming throughout the difficult political journey of the 2010s, and for that to come.

Laurie would like to thank David Adler, Simon Alcock, Myles Allen, Tom Athanasiou, John Ashton, Fernanda

Balata, Sonny Bardhan, the British Library and all the team there, Darren Baxter-Clow, Eric Beinhocker, David Bent, Dustin Benton, Grace Blakeley, Heather Boushey, Desiree Cesarini, Peter Chalkley, Ian Christie, Maeve Cohen, Jon Cracknell, Nan Craig, Edward Davey, Michael Davies, Joan Diamond, Joshua Emden, Miatta Fahnbulleh, Andrew Fanning, Anna Fielding, Thomas Fricke, the Friends Provident Foundation, Clare Gerada, Rebecca Gibbs, Abigail Gibson, Fiona Godley, Anya Gopfert, Kleoniki Gounaris, Matthew Green, Joe Guinan, Jörg Haas, Leslie Harroun, Robin Harvey, Tom Hill, Cameron Hepburn, Paul Hoggett, Leo Hollis, Christian Holz, Maria van Hove, Abi Hynes, William Hynes, Lara Iannelli, Michael Jacobs, Sofie Jenkinson, Antonia Jennings, Leonie Jordan, Astrid Kann-Rasmussen, Terry Kemple, Neil Kinnock, Oliver Kirby, Richard Kozul-Wright, Irene Krarup, Haniell Langton-Laybourn, Clive Lewis, Simon Lewis, Peter Lipman, Hywel Lloyd, Patrick Love, Linda Luxon, Mark Lynas, Laurie Macfarlane, Mark Maslin, Asher Miller, Joe Mitchell, George Monbiot, Hannah Münch, Luke Murphy, Leo Murray, Hettie O'Brien, the Omidyar Network, David Owen, the Partners for a New Economy, Raj Patel, David Pencheon, Rick van der Ploeg, Dave Powell, Harry Quilter-Pinner, Public Service Broadcasting, Carola Rackete, Lesley Rankin, Asad Rehman, Carys Roberts, Aaron Robertson, Johan Rockström, Peter Roderick, Adrian Shannon, Robert Skidelsky, Brian Valbjørn Sørensen, Achim Steiner, Julia Steinberger, Robin Stott, Nick Taylor, Alex Teytelboym, Simon Tilford, Todd Tucker, Olivia Vaughan, Danielle

Walker-Palmer, Bob Ward, Paweł Wargan, Bob Watson, David Wastell, Jonathan Watts, Nick Watts, David Wearing, Paddy West, Farhana Yamin, Yuan Yang, as well as all those who attended the programme of research roundtables for the project – Responding to Environmental Breakdown – he ran at IPPR, and the many other people whose thoughts and support have fed into this book. Above all, thanks go to Andrew, Sherren and Stephanie, whose love and support remind us of what we are fighting for.

Mat would like to thank David Adler, Kate Aronoff, Alyssa Battistoni, Christine Berry, Isabel Blake, Grace Blakeley, Guppi Bola, Benjamin Braun, Miriam Brett, Jedediah Britton-Purdy, Adrienne Buller, Johanna Buzowa, Aditya Chakrabortty, Catherine Colebrook, Daniel Aldana Cohen, Brett Christopher, Rosie Collington, Beka Diski and the 116th Street set, Sahil Dutta, David Edgerton, Josh Gabert-Doyon, Daniela Gabor, Eric Gade, Josh Gabert-Doyon, Joe Guinan, Jonathan Gray, Adam Greenfield, Peter Gowan, Miranda Hall, Thomas Hanna, Max Harris, Ned Hercock, Helen Hester, Arby Hisenaj, Cat Hobbs, Leo Hollis, Amelia Horgan, Fatima-Zahra Ibrahim, Michael Jacobs, Sofie Jenkinson, Antonia Jennings, Carsten Jung, Emily Kenway, Tom Kibasi, Jonty Leibowitz, Wendy Liu, Laurie Macfarlane, Rory Macqueen, Sarah Madmoud, Hannah Martin, Ewan McGaughey, Neil McInroy, Sarah McKinley, James Meadway, Keir Milburn, Ed Miliband, Leo Murray, Hettie O'Brien, Martin O'Neill, Nick Pearce, Barnaby Raine, Thea Riofrancos, Mary Robertson, Michal Rozworksi, Bertie Russell, James Schneider, Julian Siravo,

Quinn Slobodian, Sophie Smith, Nick Srnicek, Alfie Stirling, Will Stronge, Nick Taylor, Joe Todd, Adam Tooze and Hilary Wainwright, along with many, many more, for the pleasure of their conversation, thoughts and support that have in different ways fed into this book. A particular thanks to Carsten, Carys, Eric, Grace, Hettie, Laurie, Ned and Sahil for reviewing draft chapters – and especially to Adrienne, Amelia and Miriam, whose fine-reading and suggestions immeasurably improved things, and who make life at Common Wealth a pleasure. He spent what, in retrospect, seems an unlikely proportion of his childhood in the scrapyards of Johannesburg, an education in many ways on the themes of this book. For that – and for an abiding love of Mzansi and its people – he is grateful to his father, David, whose decency and kindness still inspires. To his mother, Kathy, whose strength and love has indelibly shaped me, a son's gratitude and love; without you, this book and much else would have been impossible. In Hannah, Charlie, Helen and Ben, he has always been able to rely on deep wells of generosity, love and support, for which Mathew is truly grateful. Finally, thank you to Carys, whose presence is a light and inspiration. From him, this book is dedicated to the memory of my grandmothers, Milly and Ilene, and in hope for Sydney and Innes, recent arrivals of unbridled joy.

We hope this book makes a useful contribution to the most important debate of our time: how life may once more flourish. Read it, then act, with all the love and hope in the world.

NOTES

Introduction

1 Valérie Masson-Delmotte et al., 'Global Warming of 1.5°C', *An IPCC Special Report on the Impacts of Global Warming of 1.5°C* (2018).

2 Dan Tong et al., 'Committed emissions from existing energy infrastructure jeopardize 1.5°C climate target', *Nature* 572, no. 7769 (2019): 373–7.

3 Bill McKibben, 'When it comes to climate hypocrisy, Canada's leaders have reached a new low', *Guardian*, 5 February 2020.

4 David Mackie and Jessica Murray, 'Risky Business: The Climate and the Macroeconomy', JP Morgan Economic Research, 14 January 2020.

5 James Hansen et al., 'Young people's burden: requirement of negative CO_2 emissions', *Earth System Dynamics* (2016); Jan Minx, Sabine Fuss and Gregory Nemet, 'Guest Post: Seven key things to know about "negative emissions"', *Carbon Brief*, 1 June 2018; 'Averting Climate Breakdown by Restoring Ecosystems', Call to Action, Natural Climate Solutions, accessed 22 March 2020.

6 Cynthia Scharf and Mark Turner, 'DiCaprio film heralds the era of carbon dioxide removal', Carnegie Climate Governance Initiative,

12 June 2019. Many scenarios model over 730 billion tonnes of carbon dioxide sequestered as negative emissions over the course of this century, a figure equivalent to the emissions of the US, UK, Germany and China – some of the world's biggest emitters – since the Industrial Revolution. See *IPCC Special Report*, Chapter 2, and Simon Lewis, 'Sucking carbon out of the air is no magic fix for climate emergency', *Guardian*, 1 August 2019.

7 Aude Mazoue, 'Le Pen's National Rally goes green in bid for European election votes', France24, 20 April 2019.

8 Sierra Garcia, ' "We're the virus": The pandemic is bringing out environmentalism's dark side', *Grist*, 30 March 2020.

9 Rosa Luxemburg, *The Junius Pamphlet: The Crisis in the German Social Democracy*, Young Socialist Publications, 1967.

10 Kate Aronoff, Alyssa Battistoni, Daniel Aldana Cohen, and Thea Riofrancos, *A Planet to Win: Why We Need a Green New Deal*, Verso Books, 2019.

11 Adam Tooze, 'We are living through the first economic crisis of the Anthropocene', *Guardian*, 7 May 2020.

12 William Davies, 'The holiday of exchange value', Centre for the Understanding of Sustainable Prosperity, 7 April 2020.

13 Stuart Hall and Doreen Massey, 'Interpreting the crisis', *Soundings* 44 (2010): 57–71.

14 William Davies, 'Recovering the future: The reinvention of "social law" ', *Juncture* 20, no. 3 (2013).

1. This Is about Power

1 Jacob Roggeveen, *The Voyage of Captain Don Felipe González to Easter Island*, Kraus Reprint, 1967, pp. 1–26.

2 J. Linton Palmer, 'A visit to Easter Island, or Rapa Nui, in 1868', *Journal of the Royal Geographical Society of London* 40 (1870): 167–81.

3 Michael J. Samways, *Insect Conservation: A Global Synthesis*, CABI, 2019.

4 Jared Diamond, *Collapse: How Societies Choose to Fail or Succeed*, Penguin, 2005.

5   Jared Diamond, 'Easter's end', *Discover*, 1 August 1995.

6   Paul G. Bahn and John Roger Flenley, *Easter Island, Earth Island*, Thames and Hudson, 1992.

7   Diamond, 'Easter's End'.

8   Benny Peiser, 'From genocide to ecocide: the rape of Rapa Nui', *Energy and Environment* 16, no. 3–4 (2005): 513–39.

9   Carl Friedrich Behrens, 'Another narrative of Jacob Roggeveen's visit', *The Voyage of Captain Don Felipe Gonzales to Easter Island in 1770*, Hakluyt Society, Cambridge, pp. 131–7 (Appendix 1) (1903). Quoted in Peiser, 'From genocide to ecocide'.

10 Jean-François de Galaup La Pérouse, 'A voyage round the world: In the years 1785, 1786, 1787, and 1788. Published conformably to the decree of the National Assembly of the 22d of April, 1791. In three volumes. Vol. III' (1798).

11 Terry L. Hunt and Carl P. Lipo, *The Statues that Walked: Unraveling the Mystery of Easter Island*, Simon & Schuster, 2011.

12 William Judah Thomson, *Te Pito te Henua, or Easter Island*, vol. 1, Library of Alexandria, 1891. Quoted in Peiser, 'From genocide to ecocide'.

13 Peiser, 'From genocide to ecocide'.

14 Hunt and Lipo, *The Statues that Walked*.

15 Terry L. Hunt and Carl P. Lipo, 'Late colonization of Easter Island', *Science* 311, no. 5767 (2006): 1603–06.

16 Roggeveen's journal, reproduced at easterisland.travel.

17 Clark Spencer Larsen and George Milner, *In the Wake of Contact: Biological Responses to Conquest*, Wiley-Liss, 1994.

18 K. Routledge, *The Mystery of Easter Island: The Story of an Expedition*, Sifton, Praed and Co., 1919.

19 Peiser, 'From genocide to ecocide'.

20 Alfred Métraux, *Easter Island: A Stone-age Civilization of the Pacific*, Oxford University Press, 1957, p. 47.

21 Thor Heyerdahl and Edwin N. Ferdon (eds), 'Miscellaneous papers', Forum Publishing House (distributed in US by Rand McNally), 1961, p. 76.

22 Thomas A. Püschel, Jaime Espejo, Maria-José Sanzana and Hugo A. Benítez, 'Analysing the floral elements of the lost tree of Easter Island: A morphometric comparison between the remaining ex-situ

lines of the endemic extinct species *Sophora toromiro*', *PLOS ONE* 9, no. 12 (2014): e115548.

23 Métraux, *Easter Island*, p. 38.

24 Jamil Zaki, 'Caring about tomorrow: Why haven't we stopped climate change? We're not wired to empathize with our descendants', *Washington Post*, 22 August 2019.

25 Emmanuel Macron, 'Speech of French President Emmanuel Macron at the 74th United Nations General Assembly', 24 September 2019, ambafrance.org.

26 Bill Gates, 'We didn't see this coming', *Gates Notes*, Annual Letter, 12 February 2019.

27 Simon L. Lewis and Mark A. Maslin, *Human Planet: How We Created the Anthropocene*, Yale University Press, 2018.

28 Raj Patel and Jason W. Moore, *A History of the World in Seven Cheap Things: A Guide to Capitalism, Nature, and the Future of the Planet*, University of California Press, 2017.

29 Ibid.

30 Orlando Patterson, *Slavery and Social Death: A Comparative Study*, Harvard University Press, 2018.

31 Jason W. Moore, 'World accumulation and planetary life, or, why capitalism will not survive until the "last tree is cut"', *IPPR Progressive Review* 24, no. 3 (2017): 175–202.

32 John Maynard Keynes, *The Economic Consequences of the Peace* [1919], Routledge, 2017.

33 Edmund Dene Morel, *Red Rubber: The Story of the Rubber Slave Trade Which Flourished on the Congo for Twenty Years, 1890–1910*, National Labour Press, 1919.

34 Jason W. Moore, 'The Capitalocene, Part I: On the nature and origins of our ecological crisis', *Journal of Peasant Studies* 44, no. 3 (2017): 594–630.

35 Janae Davis, Alex A. Moulton, Levi Van Sant and Brian Williams, 'Anthropocene, Capitalocene, . . . Plantationocene? A manifesto for ecological justice in an age of global crises', *Geography Compass* 13, no. 5 (2019): e12438.

2. Facing the Crisis

1 Various, *World Scientists' Warning to Humanity*, Union of Concerned Scientists, 1993.

2 Various, 'World scientists' warning to humanity: a second notice', *BioScience* 67, no. 12 (2017): 1026–8.

3 Laurie Laybourn-Langton, Lesley Rankin and Darren Baxter, 'This Is a Crisis: Facing Up to the Age of Environmental Breakdown', Institute for Public Policy Research, London, 2019.

4 Zoe Schlanger, 'Earth Overshoot Day is earlier than ever this year – and it underestimates the crisis', *Quartz*, 27 July 2019.

5 Gerardo Ceballos, Paul R. Ehrlich and Rodolfo Dirzo, 'Biological annihilation via the ongoing sixth mass extinction signaled by vertebrate population losses and declines', *Proceedings of the National Academy of Sciences* 114, no. 30 (2017): E6089–96.

6 World Wildlife Fund, 'Living Planet Report 2018: Aiming Higher', 2018.

7 Damian Carrington, 'Humanity has wiped out 60 per cent of animal populations since 1970, report finds', *Guardian*, 30 October 2018.

8 IPCC, 'Summary for policymakers', in *Climate Change and Land: An IPCC special report on climate change, desertification, land degradation, sustainable land management, food security, and greenhouse gas fluxes in terrestrial ecosystems*, 2019, in press.

9 IPBES, 'Summary for policymakers of the assessment report on land degradation and restoration of the Intergovernmental Science Policy Platform on Biodiversity and Ecosystem Services', IPBES secretariat, Bonn, 2018.

10 Ibid.

11 Thomas W. Crowther et al., 'Mapping tree density at a global scale', *Nature* 525, no. 7568 (2015): 201–5.

12 IPCC, 'Summary for policymakers'.

13 D. Cameron, C. Osborne, P. Horton and M. Sinclair, 'A sustainable model for intensive agriculture', *Grantham Centre for Sustainable Futures* 2 (2015).

14 Donald E. Canfield, Alexander N. Glazer and Paul G. Falkowski,

'The evolution and future of Earth's nitrogen cycle', *Science* 330, no. 6001 (2010): 192–6.

15 Caspar A. Hallmann et al., 'More than 75 percent decline over 27 years in total flying insect biomass in protected areas', *PLOS ONE* 12, no. 10 (2017): e0185809.

16 Jason M. Tylianakis, 'The global plight of pollinators', *Science* 339, no. 6127 (2013): 1532–3.

17 Intergovernmental Science-Policy Platform on Biodiversity and Ecosystem Services (IPBES), The Global Assessment Report on *Biodiversity and Ecosystem Services: Summary for Policymakers*, IPBES, 2019, ipbes.net.

18 Will Steffen et al., 'Trajectories of the Earth System in the Anthropocene', *Proceedings of the National Academy of Sciences* 115, no. 33 (2018): 8252–9.

19 'Planet at risk of heading towards "Hothouse Earth" state', Stockholm Resilience Centre, 2018.

20 Ryan Kelly et al., 'Recent burning of boreal forests exceeds fire regime limits of the past 10,000 years', *Proceedings of the National Academy of Sciences* 110, no. 32 (2013): 13055–60.

21 Mark Parrington, 'I think it's fair to say July Arctic Circle #wildfires are now at unprecedented levels', Twitter, 22 July 2019, @m_parrington.

22 James Temple, 'Australia's fires have pumped out more emissions than 100 nations combined', *MIT Technology Review*, 10 January 2020.

23 António Guterres, 'Secretary-general's remarks on climate change', speech on 10 September 2018, un.org.

24 The Global Food Security programme, 'Extreme weather and resilience of the global food system', Final Project Report from the UK–US Taskforce on Extreme Weather and Global Food System Resilience, 2015.

25 Chris Kent et al., 'Using climate model simulations to assess the current climate risk to maize production', *Environmental Research Letters* 12, no. 5 (2017): 054012.

26 Department of Defense, *Quadrennial Defense Review*, US Government, 2014, archive.defense.gov.

27 Laybourn-Langton, Rankin and Baxter, 'This Is a Crisis'.

28 Diane Taylor, 'Work on Caribbean island airport halted by court ruling', *Guardian*, 2 August 2018.

29 Katie Lebling, Mengpin Ge and Johannes Friedrich, '5 Charts Show How Global Emissions Have Changed Since 1850', World Resources Institute, 2 April 2018.

30 Timothy Gore, 'Extreme Carbon Inequality: Why the Paris climate deal must put the poorest, lowest emitting and most vulnerable people first', Oxfam International, 2 December 2015.

31 'National Footprint Accounts', Global Footprint Network, 2018.

32 Gore, 'Extreme Carbon Inequality'.

33 Laurie Laybourn-Langton, Joshua Embden and Lesley Rankin, 'Inheriting the Earth', IPPR, London, 2019.

34 World Health Organization, 'Gender, climate change and health', WHO, 2014.

35 Robert D. Bullard, *Environment and Morality: Confronting Environmental Racism in the United States*, United Nations Research Institute for Social Development, 2004.

36 'World faces "climate apartheid" risk, 120 more million in poverty: UN expert', UN News, 25 June 2019.

37 Daniel Politi, 'El Paso suspect reportedly a Trump supporter who wrote racist, anti-immigrant manifesto', *Slate*, 3 August 2019.

38 IPBES, 'Summary for policymakers of the assessment report'.

39 Delphine Strauss, 'French "gilets jaunes" show pain of Macron's tax policy', *Financial Times*, 4 December 2018.

40 Aude Mazoue, 'Le Pen's National Rally goes green in bid for European election votes', France24, 20 April 2019.

41 Ibid.

42 Luke Darby, 'What is eco-fascism, the ideology behind attacks in El Paso and Christchurch?', *GQ*, 7 August 2019.

43 See, for example: Gie Vanlommel, 'Alt-right start stickercampagne', Twitter, 27 March 2019, @Gievanlommel.

44 Kanta Kumari Rigaud et al., 'Groundswell', World Bank (2018).

45 Stella Schaller and Alexander Carius, *Convenient Truths: Mapping Climate Agendas of Right-Wing Populist Parties in Europe*, Adelphi, 2019.

3. Beyond the Ruins

1 Katharina Pistor, *The Code of Capital: How the Law Creates Wealth and Inequality*, Princeton University Press, 2019.

2 Laurie Laybourn-Langton and Michael Jacobs, 'Paradigm shifts in economic theory and policy', *Intereconomics* 53, no. 3 (2018): 113–18.

3 Kate Raworth, 'Doing the Doughnut at the G20?', 1 December 2018, kateraworth.com.

4 Quinn Slobodian, *Globalists: The End of Empire and the Birth of Neoliberalism*, Harvard University Press, 2018; Wendy Brown, *Undoing the Demos: Neoliberalism's Stealth Revolution*, Zone / Near Futures, 2015; William Davies, *The Limits of Neoliberalism: Authority, Sovereignty and the Logic of Competition*, SAGE Publications, 2014.

5 Oxfam, '5 shocking facts about extreme global inequality and how to even it up', 2020, oxfam.org.

6 William Callison and Zachary Manfredi (eds), *Mutant Neoliberalism: Market Rule and Political Rupture*, Fordham University Press, 2019.

7 George Monbiot, *Out of the Wreckage: A New Politics for an Age of Crisis*, Verso Books, 2017.

8 Alfie Stirling and Laurie Laybourn-Langton, 'Time for a new paradigm? Past and present transitions in economic policy', *Political Quarterly* 88, no. 4 (2017): 558–69.

9 Raymond Williams, *Resources of Hope: Culture, Democracy, Socialism*, Verso Books, 1989.

10 Sanjukta Paul, 'Antitrust as allocator of coordination rights', *UCLA Law Review* 67, no. 2, (2020): 4–63.

11 This section draws on the work of Erik Olin Wright in particular, especially *Envisioning Real Utopias*, Verso Books, 2010 and *How to Be an Anticapitalist in the Twenty-First Century*, Verso Books, 2019.

12 Ibid.

13 Lesjek Kolakowski, quoted in Peter Clarke, 'Seeing it all', *London Review of Books* 11, no. 19, 12 October 1989.

14 Wright, *How to be an Anticapitalist*.

4. After Empire

1 Susan E. Strahan and Anne R. Douglass, 'Decline in Antarctic ozone depletion and lower stratospheric chlorine determined from Aura Microwave Limb Sounder observations', *Geophysical Research Letters* 45, no. 1 (2018): 382–90.

2 Kate Driscoll Derickson, 'Resilience is not enough', *City* 20, no. 1 (2016): 161–6.

3 Centre for Local Economic Strategies (CLES) and Preston City Council, 'How we built community wealth in Preston', 2019, cles.org.

4 Ibid.

5 Benzamin Yi, 'Infographic: The Cleveland Model', Democracy Collaborative, 12 September 2014.

6 The Health Foundation, 'The NHS as an anchor institution', 2019, health.org.uk.

7 Office of Fair Trading, *The Pharmaceutical Price Regulation Scheme*, 2007, nationalarchives.gov.uk.

8 Public Health England and NHS England, *Reducing the Use of Natural Resources in Health and Social Care – 2018 report*, 2018.

9 Ibid.

10 Mark Curtis and Tim Jones, 'Honest Accounts 2017: How the world profits from Africa's wealth', Global Justice Now, 2017, jubileedebt. org.uk.

11 Christian Aid and Jubilee Debt Campaign, *The New Global Debt Crisis*, May 2019.

12 Gabriel Zucman, *The Hidden Wealth of Nations: The Scourge of Tax Havens*, University of Chicago Press, 2015; Tax Justice Network, 'Tax Avoidance and Evasion: The Scale of the Problem', 2017.

13 United Nations Conference on Trade and Development (UNCTAD), *The Covid-19 Shock to Developing Countries*, UN, March 2020.

14 'UN poverty expert warns against tsunami of unchecked privatisation', Office of the High Commissioner for Human Rights, 19 October 2018.

15 P. Eberhardt, C. Olivet and L. Steinfort, *One Treaty to Rule Them All*, Corporate Europe Observatory and the Transnational Institute, 2018.

16 Yamina Sahed, *The Energy Charter Treaty: Assessing Its Geopolitical, Climate and Financial Impacts*, OpenEXP, September 2019.

17 Laurie Laybourn-Langton and Lesley Rankin, 'Our Responsibility', IPPR, 2019.

18 Ibid.

5. Financing Plenty

1 Timothy Mitchell, *Carbon Democracy: Political Power in the Age of Oil*, Verso Books, 2011.

2 Adam Tooze, 'How coronavirus almost brought down the global financial system', *Guardian*, 14 April 2020.

3 Ibid.

4 J. W. Mason, 'Socialise finance', *Jacobin*, 28 November 2016.

5 'Banking on Climate Change: Fossil Fuel Finance Report Card 2019', BankTrack, 2019.

6 Richard Bridle, Shruti Sharma, Mostafa Mostafa and Anna Geddes, 'Fossil Fuel to Clean Energy Subsidy Swaps: How to pay for an energy revolution', Global Subsidies Initiative report, 2019, iisd.org.

7 Alan Livsey, 'Lex in depth: the $900bn cost of "stranded energy assets"', *Financial Times*, 4 February 2020.

8 Adam Tooze, 'Why central banks need to step up on global warming', *Foreign Policy*, 20 July 2019.

9 Patrick Greenfield, 'World's top three asset managers oversee $300bn fossil fuel investments', *Guardian*, 12 October 2019.

10 Richard Partington, 'Bank of England boss says global finance is funding 4C temperature rise', *Guardian*, 15 October 2019.

11 Patrick Greenfield and Jonathan Watts, 'JP Morgan economists warn climate crisis is threat to human race', *Guardian*, 21 February 2020.

12 ESG funds are portfolios primarily of equities and/or bonds which have integrated environmental, social and governance factors into the investment process.

13 Dale Jackson, '"Greenwashing" in ETFs: Why some socially responsible funds may be misleading investors', *Globe and Mail*, 5 November 2019.

14 Daniela Gabor, 'Why shadow banking is bigger than ever', *Jacobin*, 27 November 2018.

15 Stefan Avdjiev, Mary Everett, Philip R. Lane and Hyun Song Shin, 'Tracking the international footprints of global firms', *BIS Quarterly Review* (2018): 47–66; Ann Pettifor, *The Case for the Green New Deal*, Verso Books, 2019.

16 Joseph Noss and Rhiannon Sowerbutts, 'The implicit subsidy of banks', *Bank of England Financial Stability Paper No. 15*, May 2012.

17 Thea Riofrancos, 'Plan, mood, battlefield – reflections on the Green New Deal', *Viewpoint Magazine*, 16 May 2019.

18 Hyman Minsky, *Stabilizing an Unstable Economy*, McGraw-Hill Professional, 1986.

19 Giovanni Arrighi, *The Long Twentieth Century: Money, Power and the Origins of Our Times*, Verso Books, 2010.

20 Gerald A. Epstein, 'Introduction: Financialization and the World Economy' in Gerald A. Esptein (ed.), *Financialization and the World Economy*, Edward Elgar Publishing, 2006.

21 Adair Turner, *Between Debt and the Devil: Money, Credit, and Fixing Global Finance*, Princeton University Press, 2015.

22 Emmanuel Saez and Gabriel Zucman, *The Triumph of Injustice: How the Rich Dodge Taxes and How to Make Them Pay*, W. W. Norton & Company, 2019.

23 UNCTAD, *Trade and Development Report 2019: Financing a Global Green New Deal*, 2019.

24 World Bank, *10 Years of Green Bonds: Creating the Blueprint for Sustainability Across Capital Markets*, World Bank, 18 March 2019.

25 Keston Perry, *Realising Climate Reparations: Towards a Global Climate Stabilization Fund and Resilience Fund Programme for Loss and Damage in Marginalised and Former Colonised Societies*, United Nations Association of the United Kingdom, 2020.

26 William Davies, 'Recovering the future: The reinvention of "social law"', *Juncture* 20, no. 3 (2013): 216–22.

27 Billy Nauman, 'ESG money market funds grow 15 per cent in first half of 2019', *Financial Times*, 14 July 2019.

28 Dirk Bezemer, Josh Ryan-Collins, Frank van Lerven and Lu Zhang, 'Credit Where It's Due: A Historical, Theoretical and Empirical Review of Credit Guidance Policies in the 20th Century', Working

Paper IIPP WP 2018-11, UCL Institute for Innovation and Public Purpose, December 2018.

29 Ibid; Turner, *Between Debt and the Devil*.

30 Laurie Macfarlane, 'A spectre is haunting the West – the spectre of authoritarian capitalism', openDemocracy, 2020.

31 Dirk Bezemer et al., 'Credit where it's due'.

32 Tooze, 'Why central banks need to step up on global warming'.

33 Ann Pettifor, *Just Money: How Society Can Break the Despotic Power of Finance*, Commonwealth Publishing, 2015.

34 Benjamin Braun and Leah Downey, 'Against Amnesia: Re-Imagining Central Banking', Council on Economic Policies, January 2020.

35 Ibid.

36 Maximilian Krahé, 'The dog that didn't bark: Inflation and power in the contemporary capitalist state', *Renewal* 28, no. 1 (2020): 72–83.

37 Alfie Stirling, Dave Powell and Frank van Lerven, 'Public Finance for a Green New Deal: Why We Need to Change the Rules', Common Wealth, 2019.

38 Committee on Climate Change, 'Net Zero: The UK's Contribution to Stopping Global Warming', 2019.

39 Alfie Stirling et al., 'Public Finance for a Green New Deal'.

40 Ibid.

41 J. W. Mason, 'The Macroeconomic Case for a Green New Deal', Roosevelt Institute, 2019.

42 Christine Berry and Laurie Macfarlane, *A New Public Banking Ecosystem*, report to the Labour Party commissioned by the Communication Workers Union and The Democracy Collaborative, 2019.

43 Davies, 'Recovering the future'.

44 Christine Berry and Joe Guinan, *People Get Ready! Preparing for a Corbyn Government*, OR Books, 2019.

45 Ben Wray, 'De-financialising the economy: What is to be done?', openDemocracy, 10 October 2019.

46 Tooze, 'Why central banks need to step up on global warming'.

47 Thomas M. Hanna, *The Crisis Next Time: Planning for Public Ownership as an Alternative to Corporate Bank Bailouts*, Next System Project, 2018.

6. Owning the Future

1 Gordon H. Chang and Shelley Fisher Fishkin (eds), *The Chinese and the Iron Road: Building the Transcontinental Railroad*, Stanford University Press, 2019.

2 Richard White, *Railroaded: The Transcontinentals and the Making of Modern America*, W. W. Norton & Company, 2012.

3 Timothy Mitchell, *Carbon Democracy: Political Power in the Age of Oil*, Verso Books, 2011.

4 Timothy Mitchell, 'A Brief History of Capitalization: From Colonialism to Life Itself', public lecture, University of California, Irvine, 23 April 2015.

5 Bill Lazonick, 'Stock buybacks: From retain-and-reinvest to down-size-and-distribute', *Brookings*, 17 April 2015.

6 Laura Horn, 'The Financialization of the Corporation', in Grietje Baars and André Spicer (eds), *The Corporation*, Cambridge University Press, 2017, pp. 281–90.

7 James Gard, 'Record 2019 for UK dividends', *Morningstar*, 27 January 2020.

8 Alfie Stirling, 'Time for Demand', NEF, 2019.

9 William Lazonick, Mustafa Erdem Sakinç and Matt Hopkins, 'Why stock buybacks are dangerous for the economy', *Harvard Business Review*, 7 January 2020.

10 Thomas M. Hanna, *Our Common Wealth: The Return of Public Ownership in the United States*, Manchester University Press, 2018.

11 Wendy Brown, *Undoing the Demos: Neoliberalism's Stealth Revolution*, Zone Books, 2017.

12 Erik Olin Wright, *How to Be an Anticapitalist in the Twenty-First Century*, Verso Books, 2019. Joe Guinan and Martin O'Neill, 'The institutional turn: Labour's new political economy', *Renewal* 26, no. 2 (2018): 5–16.

13 Ibid.

14 Quinn Slobodian, *Globalists: The End of Empire and the Birth of Neoliberalism*, Harvard University Press, 2018.

15 Karl Polanyi, *The Great Transformation: The Political and Economic Origins of Our Time*, Beacon Press, 2011 (reissued).

16 Sanjukta Paul, 'Antitrust as allocator of coordination rights', *UCLA Law Review* 67, no. 2, (2020): 4–63.

17 Jedediah Britton-Purdy et al., 'Building a Law-Political-Economy Framework: Beyond the Twentieth-Century Synthesis', *Yale Law Journal*, April 2020.

18 Ferreras, *Firms as Political Entities*.

19 Patrick Ireland, 'Company law and the myth of shareholder ownership', *Modern Law Review* 62, no. 1 (1999): 32–57.

20 Simon Deakin, 'The corporation as commons: rethinking property rights, governance and sustainability in the business enterprise', *Queen's Law Journal* 37, no. 2 (2011): 339–381.

21 Carolyn Sissoko, 'The problem with shareholder bailouts isn't moral hazard, but undermining state capacity', *Just Money*, 10 April 2020.

22 Simon Deakin, 'The Corporation as Commons'.

23 Kate Aronoff, Alyssa Battistoni, Daniel Cohen and Thea Riofrancos, *A Planet to Win: Why We Need a Green New Deal*, Verso Books, 2019.

24 Ferreras, *Firms as Political Entities*.

25 Ewan McGaughey, *Principles of Enterprise Law*, Cambridge University Press, forthcoming.

26 Katharina Pistor, *The Code of Capital: How the Law Creates Wealth and Inequality*, Princeton University Press, 2019.

27 Jedediah Britton-Purdy et al., 'Building a Law-and-Political-Economy Framework: Beyond the Twentieth-Century Synthesis', Columbia Public Law Research Paper, no. 14–657 (2020).

28 Ewan McGaughey, 'Ending shareholder monopoly: why workers' votes promote good corporate governance', LSE Politics Blog, 30 November 2017.

29 Robin Blackburn, *Banking on Death, or, Investing in Life: The History and Future of Pensions*, Verso Books, 2003.

30 David H. Webber and Michael A. McCarthy, 'Is Labor's Future in Labor's Capital? A Debate', *Law and Political Economy* blog, 22 June 2019.

31 Marjorie Kelly and Ted Howard, *The Making of a Democratic Economy: How to Build Prosperity for the Many, Not the Few*, Berrett-Koehler Publishers, 2019.

32 Thomas M. Hanna and Andrew Cumbers, 'Democratic ownership: A primer', openDemocracy, 11 March 2019.

33 Branko Milanovic, *Capitalism, Alone: The Future of the System That Rules the World*, Harvard University Press, 2019.

34 Matt Bruenig, 'Bernie Sanders gives the nod to funds socialism', People's Policy Project, 29 May 2019.

35 Joe Guinan, 'Socialising capital: Looking back on the Meidner Plan', *International Journal of Public Policy* 15, no. 1 (2019): 38–58.

36 Eric Lonergan and Mark Blyth, 'Beyond Bailouts', IPPR, 2020.

37 Grace Blakeley, *Stolen: How to Save the World from Financialisation*, Repeater Books, 2019.

38 Ibid.

39 Thomas M. Hanna and Joe Guinan, 'Democratising capital at scale: cooperative enterprise and beyond', openDemocracy, 7 August 2013.

40 Erik Olin Wright, *Envisioning Real Utopias*, Verso Books, 2010.

41 Roberto Unger, *The Left Alternative*, Verso Books, 2009.

42 David Edgerton, *The Rise and Fall of the British Nation: A Twentieth-Century History*, Allen Lane, 2018.

43 Satoko Kishimoto, Olivier Petitjean and Lavinia Steinfort, *Reclaiming Public Services: How Cities and Citizens Are Turning Back Privatisation*, Transnational Institute, 2017.

44 Johanna Bozuwa and Carla Skandier, 'Shifting Ownership for the Energy Transition in the Green New Deal: A Transatlantic Proposal', Common Wealth, 2019.

7. Commoning the Earth

1 Massimo Libardi, 'Leaked documents show Brazil's Bolsonaro has grave plans for the Amazon rainforest', openDemocracy, 21 August 2019.

2 Jonathan Watts, 'G7 can't turn a blind eye to ecocide in the Amazon', *Observer*, 25 August 2019.

3 Richard Seymour, 'Bolsonaro', Patreon, 30 October 2018.

4 Jean Wyllys, 'Fascism in Brazil thrives on creating enemies', openDemocracy, 30 July 2019.

5 Richard Seymour, 'Misanthropocene', Patreon, 22 March 2019.

6 Dom Phillips, 'Amazon rainforest "close to irreversible tipping point" ', *Guardian*, 23 October 2019.

7 John Bellamy Foster, Brett Clark and Richard York, *The Ecological*

Rift: Capitalism's War on the Earth, Monthly Review Press, 2010.

8 Christophe Bonneuil and Jean-Baptiste Fressoz, *The Shock of the Anthropocene: The Earth, History and Us*, Verso Books, 2017.

9 James Ward et al., 'Is Decoupling GDP Growth from Environmental Impact Possible?', *PLOS ONE* 11, no. 10 (2016): e0164733.

10 John Bellamy Foster, *Marx's Ecology: Materialism and Nature*, Monthly Review Press, 2000.

11 Andreas Malm, *Fossil Capital: The Rise of Steam Power and the Roots of Global Warming*, Verso Books, 2016.

12 Karl Marx, *Capital: A Critique of Political Economy, Vol. 1* [1867], Penguin Classics, 1990.

13 Richard Seymour, 'Do we even have the energy for capitalism?', Patreon, 27 September 2017.

14 Ben Tarnoff, 'The data is ours!' *Logic* 4, 1 April 2018.

15 Clare Coffey et al., *Time to Care: Unpaid and Underpaid Care Work and the Global Inequality Crisis*, Oxfam, 2020.

16 Ibid.

17 Alyssa Battistoni, 'Living together shouldn't put us at war with one another or with the earth: an interview with Jedediah Purdy', *Jacobin*, 3 October 2019.

18 Raj Patel and Jason W. Moore, *A History of the World in Seven Cheap Things: A Guide to Capitalism, Nature, and the Future of the Planet*, Verso Books, 2017.

19 Elinor Ostrom, *Governing the Commons: The Evolution of Institutions for Collective Action*, Cambridge University Press, 1983.

20 Álvaro Sevilla-Buitrago, 'Capitalist Formations of enclosure: space and the extinction of the commons', *Antipode*, 47, no. 4 (2015): 999–1020.

21 Marx, *Capital*.

22 Silvia Federici, *Caliban and the Witch: Women, the Body and Primitive Accumulation*, Autonomedia, 2017.

23 David Harvey, *The New Imperialism*, Oxford University Press, 2003.

24 Michel Bauwens, 'Michel Bauwens on the Commons Transition', *The Commons Transition*, 21 December 2016.

25 Ostrom, *Governing the Commons*.

26 Danijela Dolenec, 'Socialism and the Commons', openDemocracy, 31 July 2013.

27 Jeremy Gilbert, *Common Ground Democracy and Collectivity in an Age of Individualism*, Pluto Press, 2013.

28 Keir Milburn and Bertie Russell, 'Public-Common Partnerships: Building New Circuits of Collective Ownership', Common Wealth, 2019.

29 Silvia Federici, *Re-enchanting the World: Feminism and the Politics of the Commons*, PM Press, 2018.

30 Nick Estes, 'A Red Deal', *Jacobin*, 6 August 2019.

31 Derek Wall, *The Commons in History: Culture, Conflict, and Ecology*, MIT Press, 2014.

32 Brett Christophers, *The New Enclosure: The Appropriation of Public Land in Neoliberal Britain*, Verso Books, 2018.

33 Guy Shrubsole, *Who Owns England?: How We Lost Our Green and Pleasant Land, and How to Take it Back*, William Collins, 2019.

34 J. W. Mason, 'Posts in three lines, coronavirus edition', 21 April 2020, jwmason.org.

35 The Trust would purchase land for individuals, councils or cooperatives seeking help to purchase housing; the individual, council or coop would only buy the bricks and mortar. See Beth Stratford and Duncan McCann, 'The Common Ground Trust: A Route out of the Housing Crisis', Progressive Economics Group, 2018.

36 Josh Ryan-Collins, Toby Lloyd and Laurie Macfarlane, *Rethinking the Economics of Land and Housing*, ZED Books, 2017.

37 Daniel Aldana-Cohen, 'A Green New Deal for Housing', *Jacobin*, 8 February 2019.

38 Shiying Wang and Mengpin Ge, 'Everything You Need to Know About the Fastest-Growing Source of Global Emissions: Transport', World Resources Institute, 2019.

39 Yannick Oswald, Anne Owen and Julia K. Steinberger, 'Large inequality in international and intranational energy footprints between income groups and across consumption categories', *Nature Energy* 5 (2020): 231–9.

40 Thea Riofrancos, 'What green costs', *Nature* 9, (2019).

41 Leo Murray, 'Away With All Cars', Common Wealth, 2019.

42 Leo Murray and Jamie Beevo, 'Jet, Set, Go: The Case for Electric-Only UK Private Jet Flights from 2025', Common Wealth, 2019.

43   Timothy Mitchell, *Carbon Democracy: Political Power in the Age of Oil*, Verso Books, 2011.

44   Zoe Schlanger, 'If shipping were a country, it would be the world's sixth-biggest greenhouse gas emitter', World Economic Forum, 18 April 2018.

45   Ashley Dawson, *Extreme Cities: The Peril and Promise of Urban Life in the Age of Climate Change*, Verso Books, 2017.

46   Aldana Cohen, 'A Green New Deal for Housing'.

47   Simon Lewis and Mark Maslin, *The Human Planet: How We Created the Anthropocene*, Pelican Books, 2018.

48   Cheikh Mbow, Cynthia Rosenzweig et al., 'Food Security in Climate Change and Land: An IPCC special report on climate change, desertification, land degradation, sustainable land management, food security, and greenhouse gas fluxes in terrestrial ecosystems', IPCC, 2019.

49   Richard Seymour, 'How food insecurity keeps the workforce cowed', *Guardian*, 23 August 2012.

50   'Global Agriculture towards 2050', Food and Agriculture Organization of the United Nations, 2009.

51   Patel and Moore, *A History of the World in Seven Cheap Things*.

52   Jason Moore, 'World Accumulation and Planetary Life, or Why Capitalism Will Not Survive Until the "Last Tree Is Cut" ', speech to Political Economy Research Centre/Centre for the Understanding of Sustainable Prosperity, 10 October 2017.

53   Baher Kamal, 'One third of food lost, wasted – enough to feed all hungry people', Inter Press Service, 28 November 2017.

54   'Mike Davis on pandemics, super-capitalism, and the struggles of tomorrow', *Mada Masr*, 30 March 2020.

55   Raj Patel and Jim Goodman, 'A Green New Deal for Agriculture', *Jacobin*, 4 April 2019.

56   'Key Statistics and Trends in Trade Policy 2018', United Nations Conference on Trade and Development, 2019.

57   Tad Friend, 'Can a burger help solve climate change?' *New Yorker*, 23 September 2019.

58   Adam Greenfield, *Radical Technologies: The Design of Everyday Life*, Verso Books, 2017.

59   David Edgerton, *The Shock of the Old: Technology and Global History Since 1900*, Oxford University Press, 2006.

60　Shoshana Zuboff, *The Age of Surveillance Capitalism: The Fight for a Human Future at the New Frontier of Power*, Profile Books, 2019.

61　The Editors, 'Go ask Alice', *Logic* 4, 1 April 2018.

62　Tarnoff, 'The data is ours!'

63　Kenya Evelyn, 'Amazon CEO Jeff Bezos grows fortune by \$24bn amid coronavirus pandemic', *Guardian*, 15 April 2020.

64　Greenfield, *Radical Technologies*.

65　Jedediah Britton-Purdy, *This Land Is Our Land: The Struggle for a New Commonwealth*, Princeton University Press, 2019.

66　Miranda Hall, Mathew Lawrence, Sara Mahmoud and Adrienne Buller, 'Full Fibre Futures: Democratic Ownership and the UK's Digital Infrastructure', Common Wealth, 2019.

67　Zarine Kharazian, '"Technology is neither good, nor bad; nor is it neutral": the case of algorithmic biasing', *SSRMC*, 18 November 2016.

68　Tarnoff, 'The data is ours!'

69　Greenfield, *Radical Technologies*.

70　Wendy Liu, *Abolish Silicon Valley: How to Liberate Technology from Capitalism*, Watkins Media, 2020.

71　Mathew Lawrence and Laurie Laybourn-Langton, 'The Digital Commonwealth: From Private Enclosure to Collective Benefit', IPPR, 2018.

72　Jonathan Gray, 'Is "another internet possible"? Inside Labour's digital infrastructure', openDemocracy, 2019.

73　Open Data Institute, 'Data Trusts: Lessons from Three Pilots', ODI, 2019

74　Tarnoff, 'The data is ours!'

75　Bruno Latour, 'Some Advantages of the notion of "critical zone" for geopolitics', *ScienceDirect* 10 (2014): 3–6.

76　Evgeny Morozov, 'Digital socialism', *New Left Review* 116, Mar/June 2019.

77　Evgeny Morozov, 'Socialise the data centres', *New Left Review*, 91, Jan/Feb 2015.

78　Richard Seymour, *The Twittering Machine*, Indigo Press, 2019.

79　Francesca Bria and Evgeny Morozov, 'Rethinking the Smart City: Democratizing Urban Technology', Rosa Luxemburg Stiftung, January 2018.

80 Jeremy Gilbert, *Twenty-First-Century Socialism*, Polity Press, 2020.

81 Tom Mills and Dan Hind, 'Public ownership of the public sphere', *New Socialist*, 4 March 2018.

82 Dan Hind, 'The British Digital Cooperative: A New Model Public Sector Institution', *Common Wealth*, 2019.

83 Ibid.

84 Tom Mills and Dan Hind, 'Public ownership of the public sphere', *New Socialist*, 4 March 2018.

85 This analysis draws on Duncan McAnn, 'Commoning IP', Common Wealth, 2020.

86 Duncan McCann, 'Power and Accountability in the Digital Economy', NEF, 2019.

87 Sharif Abdel Kouddous, quoting Mike Davis, in 'Mike Davis on pandemics, super-capitalism and the struggles of tomorrow', *Portside*, 30 March 2020.

8. Thriving Not Surviving

1 Adrian Smith, 'The Lucas Plan: What can it tell us about democratising technology today?', *Guardian*, 22 January 2014.

2 Hilary Wainwright and Dave Elliott, *The Lucas Plan: New Trade Unionism in the Making*, Allison & Busby, 1981.

3 Smith, 'The Lucas Plan'.

4 William Davies, 'The holiday of exchange value', Centre for Understanding of Sustainable Prosperity, 7 April 2020.

5 Anna Lowenhaupt Tsing, *The Mushroom at the End of the World: On the Possibility of Life in Capitalist Ruins*, Princeton University Press, 2015.

6 Kathi Weeks, *The Problem with Work: Feminism, Marxism, Antiwork Politics, and Postwork Imaginaries*, Duke University Press, 2011.

7 Linda McDowell, Adina Batnitzky and Sarah Dyer, 'Division, segmentation, and interpellation: the embodied labours of migrant workers in a Greater London hotel', *Economic Geography* 83, no. 1 (2015): 1–25.

8 Genevieve LeBaron (ed.), *Researching Forced Labour in the Global Economy: Methodological Challenges and Advances*, Oxford University Press, 2018.

9 Nancy Fraser, 'Contradictions of capital and care', *New Left Review* 100, July/August 2016.

10 Pablo Gilabert and Martin O'Neill, 'Socialism', *Stanford Encyclopedia of Philosophy*, 2019.

11 Karl Marx, *Capital: A Critique of Political Economy, Vol. 1* [1867], Penguin Classics, 1990.

12 Elizabeth Anderson, *Private Government: How Employers Rule Our Lives (and Why We Don't Talk About It)*, Princeton University Press, 2017.

13 Isabelle Ferreras, *Firms as Political Entities: Saving Democracy through Economic Bicameralism*, Cambridge, 2017.

14 Tithi Bhattacharya, 'Three ways a Green New Deal can promote life over capital', *Jacobin*, 10 June 2019.

15 Anton Jäger, 'Why "post-work" doesn't work', *Jacobin*, 19 November 2018.

16 Erik Olin Wright, *Envisioning Real Utopias*, Verso Books, 2010.

17 Gilabert and O'Neill, 'Socialism'.

18 Alyssa Battistoni, 'Living, not just surviving', *Jacobin*, 15 August 2017.

19 Erik Olin Wright, 'Compass points', *New Left Review*, Sept/Oct 2006.

20 Naomi Klein, *This Changes Everything: Capitalism vs Climate Change*, Simon & Schuster, 2014.

21 Bhattacharya, 'Three ways a Green New Deal can promote life over capital'.

22 Battistoni, 'Living, Not Just Surviving'.

23 Nick Srnicek and Helen Hester, 'The crisis of social reproduction and the end of work', in *The Age of Perplexity: Rethinking the World We Knew*, BBVA, OpenMind, Penguin Random House Grupo Editorial, 2017.

24 Ibid.

25 Ibid.

26 Salar Mohandesi and Emma Teitelman, 'Without Reserves', in Tithi Bhattacharya (ed.), *Social Reproduction Theory: Remapping Class, Recentring Oppression*, Pluto Press, 2017.

27 Ibid.

28 Camille Barbagallo, *The Political Economy of Reproduction:*

Motherhood, Work and the Home in Neoliberal Britain, PhD Thesis, University of East London, 2016.

29 Helen Hester, 'The World Transformed Through Care', IPPR, 10 October 2017.

30 The Lancet, *The Lancet Commission on Health and Climate Change*, 2015.

31 Karel Williams et al., 'Manifesto for the foundational economy', Centre of Research on Socio-Cultural Change, Working Paper no. 131, 2013.

32 Ibid.

33 Sophie Lewis, *Full Surrogacy Now: Feminism Against Family*, Verso Books, 2019.

34 Leigh Phillips and Michal Rozworski, *The People's Republic of Walmart: How the World's Biggest Corporations are Laying the Foundation for Socialism*, Verso Books, 2019.

35 Dirk Bezemer, Josh Ryan-Collins, Frank van Lerven and Lu Zhang, 'Credit Where It's Due: A Historical, Theoretical and Empirical Review of Credit Guidance Policies in the Twentieth Century', Working Paper IIPP WP 2018-11, UCL Institute for Innovation and Public Purpose, December 2018.

36 Aronoff et al., *A Planet to Win*.

37 David Autor and Anna Salomons, 'Is Automation Labour-Displacing? Productivity Growth, Employment and the Labour Share, Brookings Papers, 27 February 2018.

38 Mathew Lawrence, Carys Roberts and Loren King, 'Managing Automation: Employment, Inequality and Ethics in the Digital Age', IPPR, 2017.

39 Roberto Unger, *The Religion of the Future*, Harvard University Press, 2014.

40 Nick Srnicek, *Platform Capitalism*, Polity Books, 2016.

41 Evgeny Morozov, 'Digital socialism', *New Left Review* 116, March–June 2019.

42 Richard M. Locke, *The Promise and Limits of Private Power: Promoting Labour Standards in a Global Economy*, Cambridge University Press, 2013.

43 Thomas M. Hanna and Mathew Lawrence, 'Democratic Public Ownership: The New Frontiers', Common Wealth, 2020.

44 Hilary Wainwright, *Reclaim the State: Experiments in Popular Democracy*, Seagull Books, 2009.

45 Colin Crouch, *Post-Democracy*, Polity Press, 2004.